# Contents

# Preface

Since the first edition of *Practical Photovoltaics* was published, both the alternative energy field and the photovoltaics industry have experienced considerable change. While the pressing need for energy conservation and alternative energies continues, the panic has abated and government involvement in research and development has declined drastically. [Note to Edition 3.1: In 2001, the panic returned.] Nonetheless, the photovoltaics industry has become international in scope with an existence that goes far beyond the programs set up by the Department of Energy or other government agencies. More solar cell companies have started up, making photovoltaic cells by new processes, while the original companies have evolved and grown.

The amorphous silicon solar cell has come into its own, powering millions of calculators and other small devices; polycrystalline silicon blocks are being made into solar cells in Germany and Japan, as well as in the United States. Sales of solar cell modules and systems are growing fast, increasing from about five megawatts per year when the first edition was published in 1981 to better than 200 megawatts a year now, reflecting our increasing dependence on this subtle, silent energy source.

Solar cells are now an accepted and important product, the industry is maturing, and the future looks bright. PV cells are becoming more efficient as their cost decreases. Research is continuing, not only to better understand the physics and chemistry of the present crystalline and amorphous solar cells, but also to develop entirely new cell concepts, such as high-efficiency tandem cells, and new materials, such as cadmium telluride, that are now being used in commercial products.

This third edition of *Practical Photovoltaics* has been extensively reorganized, offering an improved, even more clear presentation of its contents. Throughout, material has been brought up to date and new data added. To reflect the changes of the past few years, the chapters on "New Developments in Photovoltaic Technology" and "The Future of Photovoltaics" have been updated and expanded. The chapters on using and installing systems report recent developments and improvements. *Practical Photovoltaics, Third Edition*, is more comprehensive, useful, and current, and I look forward to its being as well received as its predecessors.

—RJK

## ACKNOWLEDGMENTS

I gratefully acknowledge the aid of Dr. Dan Trivich, Dr. Edward Wang, and others who furnished a good deal of the information on photovoltaic cells and inspired me to develop this book.

I also am happy to thank Vicky Patton, who typed the manuscript; Lawrence Komp, who drew the diagrams; David Ross Stevens, who furnished many of the workshop photographs; and the manufacturers who kindly supplied illustrations.

Finally, I wish to acknowledge the help of my workshop participants, who have assisted in the development and testing of procedures used to make practical use of photovoltaics.

# Preface
## To the First Edition

Energy independence has become a national goal. It is imperative that we gain direct control over our energy sources rather than rely on dwindling fuel supplies and uncertain political alliances. The only truly independent sources of energy are the constantly renewed resources surrounding us: the wind, flowing water, growing plants, and the sun—the ultimate energy source. More and more, we are turning to the renewable resources for our many energy needs.

Today it is relatively common to heat our homes and water with the sun. We have also developed solar techniques to dry crops, make steam for industrial processes, and even cool our buildings. But the most desirable use of the sun's energy, and the one most important to our future, is the production of electricity. Photovoltaic cells, or solar cells, can capture the sun and convert it into electricity. They are rugged, reliable, long-lasting, have no moving parts, and give off no chemical by-products. Furthermore, they can be used almost anywhere, with no damage to the environment.

Many people are interested in photovoltaic cells, but are deterred by the high price and by their lack of knowledge on how to use them. Some have the impression that solar cells are an exotic source of power best left to the space program or the distant future. This is not so. Photovoltaic cells are practical right now for many purposes and it is quite simple to learn how to use them. Anywhere electricity is needed and not available from a central utility, solar cells may be the best way to furnish that electricity.

The idea for this book grew out of my series of solar cell workshops, the first of which were held at Skyheat, a small nonprofit research foundation in Southern Indiana. Workshop participants actually assemble their own solar arrays from surplus solar cells and commonly available construction materials. It became apparent that a manual would be very useful for these ongoing workshops. This book is the result.

*Practical Photovoltaics* is more than a workshop manual or a do-it-yourself book. It serves both the technical and the nontechnical reader. The first section is devoted to a general description of solar cells and modules. Then a technical section details how to use solar cells and offers step-by-step instructions for assembling solar arrays. Chapters on how solar cells work and how they are made follow. Finally, a section on the photovoltaic industry, new scientific developments and future prospects gives an overview of this fascinating application of solar energy.

—RJK

Introduction

# The History of Solar-Generated Electricity

*John Perlin*

For over a century scientists have known that sunlight can produce electricity. The principal materials used in this process have been selenium and silicon. Though almost a century separates the use of selenium from the use of silicon to tap solar energy for electricity, discovering this capability for both happened by chance.

In the early 1870s, engineer Willoughby Smith's search for a material of high electrical resistance to test submerged undersea cables from onshore led him to try selenium. Despite its reputation for high resistance, the material failed to work as expected; investigators testing the selenium came up with divergent results. Curious as to why it performed so erratically, Smith found that the amount of light pouring onto the selenium determined its resistance. He discovered this by placing the selenium in a box with a sliding cover. When the

 From *A Golden Thread: 2500 Years of Solar Architecture and Technology* by Ken Butti and John Perlin.

cover stayed closed, the resistance of selenium rose to its highest point and remained constant. However, when Smith opened the box, the resistance dropped proportionally to the intensity of light falling onto it.

Smith's findings piqued the curiosity of two fellow English investigators, W. G. Adams and R. E. Day. During the late 1870s they subjected selenium pieces to many experiments, including one in which they passed a battery-generated current through the selenium. After several minutes, the selenium was detached from the battery and hooked up to a galvanometer, a device used to measure electrical current. The galvanometer registered a fairly intense electrical flow, but in the opposite direction of the original current. This showed, the researchers wrote, "that the passage of the battery current set up polarization in the selenium."

Adams and Day repeated this experiment with one change: after removing the selenium from the battery, they lit a candle and shined it onto the selenium. The galvanometer needle jumped as before but in the other direction. "Here there seemed to be a case of light actually producing an electromotive force within the selenium, which in this case was opposed to and could overbalance the electromotive force due to polarization," the amazed scientists observed.

The unexpected result led Adams and Day to alter their course of investigation to immediately examine "whether it would be possible to start a current in the selenium merely by the action of light." Connecting the same piece of selenium to a galvanometer the next morning, they lit a candle and placed it an inch away. Immediately, the needle reacted. Screening the selenium from the light, the needle returned to zero. After repeating the experiment in different ways and with various types of light, the investigators came up with similar results, proving they could start and maintain an electrical current in selenium solely by light. Adams and Day called the currents produced in this way "photoelectric."

Only a few years after Adams and Day discovered the first solar cell, C. E. Fritts, an American inventor, built the first solar battery. It resembled a sandwich: Fritts placed a sheet of selenium on metal backing and covered the selenium with transparent gold-leaf film. The selenium array, Fritts reported, produced a current "that is continuous, constant, and of considerable force . . . not only by exposure to sunlight, but also to dim diffused daylight, and even to lamplight."

Claiming that his selenium battery converted the sun's energy directly into electricity without fuel, Fritts aroused deep skepticism among fellow engineers and scientists. His professional contemporaries were accustomed to producing energy by burning combustibles. Any apparatus that purported to generate power without consuming matter, such as the selenium solar battery, they argued, violated laws of physics and should be summarily dismissed as fraud.

Cast under suspicion, Fritts needed to find approval in the scientific community for his novel invention to legitimize his work. So he sent a model to Werner Siemens, a German inventor and industrialist, who was one of the most respected authorities in the field of electricity at the time. To Fritts's joy, Siemens replicated the American's success and declared to the world, that "presented to us, for the first time, [is] the direct conversion of [the] energy of light into electrical energy . . . which is scientifically of the most far reaching importance."

Despite the much deserved praise for his groundbreaking work, crediting Fritts with the first conversion of sunlight into electricity was wrong. Edmond Becquerel, a French experimental physicist, deserves the honor. In 1839, Becquerel found that two different brass plates immersed in a liquid produce a continuous current when exposed to sunlight. From the admittedly crude spectral data given in Becquerel's report, it would appear that he succeeded in making a copper–cuprous oxide back-wall Schottky junction solar cell.

How Fritts's solar battery worked baffled even the great Siemens. He admitted that "in reality, we have . . . an entirely new physical phenomenon," requiring a "thorough investigation to determine what the electromotive light-action of [the] selenium depends."

Scientists did not heed Siemens's call to action until several decades later when the bold new theories of quantum mechanics and relativity legitimized the claims of people like Fritts. Scientists now pictured an electrical current as an orderly flow of electrons. In certain materials, such as selenium, the electron flow could be set in motion by directly interacting with light particles called photons: this is known as the photovoltaic effect. Studying the photovoltaic effect became an accepted area of experimentation for engineers and scientists.

Under the improved conditions for solar-electric research, scientists rediscovered the selenium solar battery and renewed humanity's dream of the world's industries humming along, fuel- and pollution-free, powered by the inexhaustible rays of the sun. Dr. Bruno Lange, a German scientist, who, in 1931, built a solar battery closely resembling Fritts's design, predicted, "In the not distant future, huge plants will employ thousands of these plates to transform sunlight into electric power . . . that can compete with hydroelectric and steam-driven generators in running factories and lighting homes." Lange's optimism and knack for PR struck a hopeful chord in the world's press. But Lange's solar battery worked no better than Fritts's, converting less than one percent of all incoming sunlight into electricity—hardly enough to justify its use as a power source. Lange's failure to deliver as promised sunk the high expectations people held and led experts like E. D. Wilson of Westinghouse Electric's photo-electricity division to judge that "the photovoltaic cell does not show promise as a power converter for solar energy."

The future of solar energy as a power source could not have looked gloomier on the eve of the discovery of the silicon solar cell, today's most commonly used material for converting sunshine into electricity. Transistor technology had raised the importance of silicon for those working in the electrical transmission field. At Bell Labs in 1954, Calvin Fuller and Gerald Pearson were experimenting with silicon rectifiers, devices that change alternating current into direct current. Fuller discovered that by adding certain impurities to silicon he could increase the rectifiers' efficiency. Pearson, an experimental physicist, began testing Fuller's new rectifiers. To his frustration, he found that he could not replicate certain measurements. They would vary according to where Pearson placed the rectifier in the lab. Suddenly, it dawned upon him that—as Willoughby Smith had discovered many years before—the changes were due to differences in lighting.

To quantify the effect that light had on the rectifier at hand, Pearson hooked it up to a voltmeter and placed the rectifier on a window sill where sunlight streamed in. The voltmeter's needle immediately jumped. Pearson called in a colleague, Morton Prince, to witness this extraordinary event. Through calculations it was found that the rectifier could convert a whooping 4% of all incoming sunlight into electricity, an exciting figure since it topped the best selenium cells by more than a factor of five.

Pearson then notified another Bell Labs colleague, Darryl Chapin, of his discovery. Bell Labs had assigned Chapin the task of finding an independent power source for the isolated repeater stations used to amplify messages on long-distance lines. Chapin had previously complained to Pearson that no solar electric converter generated enough power to do the job. The new silicon rectifier, however, seemed to be just what Chapin needed. What started out as a rectifier investigation became a solar power project. Chapin, Fuller, and Pearson joined as a team to refine what the Laboratory called the Bell Silicon Solar Battery. Several months went by and they improved the solar cell's efficiency to 6%. Ironically, there is evidence that other researchers at Bell Labs, particularly the metallurgist, R. S. Ohl, had made the same sort of discovery many years before. This was in the early 1940s when they were experimenting with growing silicon crystals and making diodes for the secret military work on radar. This work seems to have been forgotten by the mid-1950s.

The Bell Labs scientists now felt ready to take their solar cell public. What better place to unveil it than at the annual meeting of the National Academy of Science in Washington, D.C., where the specially treated silicon strips, when exposed to light, drove a toy Ferris wheel round and round and powered an FM radio that broadcast music to a packed audience.

The event made the front page. On April 26, 1954, the *New York Times* announced, "Vast Power of the Sun is Tapped by Battery Using Sand Ingredient." Under the headline, the *Times* reported that the invention of the silicon solar battery "may mark the beginning of a new era, leading eventually to the realization of one of mankind's most cherished dreams—the harnessing of the almost limitless energy of the sun for the uses of civilization."

Coincidentally, at about the same time that Chapin, Fuller, and Pearson went public with their work, Donald Reynolds and Lt. Colonel Gerald Leies of the U.S. Air Force revealed their discovery that cadmium sulfide, a yellow powder commonly found in paint, could convert sunlight directly into electricity. Processed into crystalline form, the device ran an electric clock. While the cadmium sulfide solar cell turned out to be unstable, work on this material has lead to new photovoltaic devices.

Talk that silicon solar cells might power the electrical needs of the nation forced the Bell Lab inventors to take a cautionary stance to avoid later disillusionment. Hadn't wild hopes for the selenium cells dissipated into thin air? Even after increasing the silicon solar battery's conversion ratio to a spectacular 15%, the trio avoided "making too much claim for it," according to Chapin, "because we knew it was in the laboratory stage, it was expensive, and there was much to be done before we could speak of lots of power."

Silicon solar cells were first used as the power source for a telephone relay system in an isolated rural area of Georgia. The set-up included battery storage for a nighttime power supply and worked problem-free. However, Chapin's concern about the cost of solar-generated electricity proved well-founded. After a year in operation, an economic analysis of the solar-powered telephone system showed that it was not competitive with a conventionally powered relay.

Western Electric, a subsidiary of the Bell System, licensed other companies to market products powered by silicon solar cells. One electronics firm built prototype arrays to power remote sites such as national forest look-out towers and Coast Guard buoys. The arrays did their jobs well, but no orders came. The government agencies expected small nuclear power systems to do such work in the future.

Just as the world was about to consign silicon solar cells to the curiosity heap of unique but useless inventions, the space race came along. Satellites required a long-term, autonomous power source that had to be compact and lightweight. Because conventional fuel systems and batteries were too cumbersome, those who oversaw the space race came to view silicon solar cells as the perfect answer for extraterrestrial power needs. Solar cells did not require a storage system: the sun shines 24 hours a day in outer space. They were easily the lightest, most reliable power source for space applications. Thus, the U.S. space program created the large and viable silicon solar cell industry. Starting in the late 1950s, silicon solar cells powered all of America's satellites, from the grapefruit-sized Vanguard to the giant Skylab.

Back on earth, the government rarely funded research to develop better and cheaper solar cells for ordinary commercial use, though it spent billions to develop nuclear power. Why it showed such lack of interest in photovoltaics, and such favoritism to nuclear-generated electricity, mystified many. As early as the mid-1950s, the *New York Times* suggested that the United States government "ought to transfer some of [its] interest in atomic power to solar." But Washington's attitude mirrored that of a nation hypnotized by seemingly limitless supplies of cheap fossil fuels and by the almost magical aura surrounding nuclear energy.

A few prescient individuals could see the possibility of solar cells one day changing the way the world produces its power. Harland Manchester was one of the visionaries. Countering those who belittled the role of solar cells because their terrestrial uses were confined to powering playthings, Manchester wrote, "Viewed in the light of the world's power needs these gadgets are toys. But so was the first steam-powered motor by Michael Faraday over a century ago—and it sired the whole gigantic electrical industry."

## BIBLIOGRAPHY

Adams, W. G., and R. E. Day. "On the Action of Light on Selenium." London. *(Royal Society) Proceedings* 25:115 (1877).

Fritts, C. E. "On the Fritts' Selenium Cell and Batteries." *Van Nostrand's Engineering Magazine* 32:392 (1885).

Siemens, W. "On the Electromotive Action of Illuminated Selenium Discovered by Mr. Fritts of New York." *Van Nostrand's Engineering Magazine* 32:515 (1885).

"Magic Plates Tap Sun for Power." *Popular Science Monthly* 118:41 (June 1931).

Wilson, E. D. "Power from the Sun." *Power* (October 1935):517.

Personal correspondence with Daryl Chapin.

*New York Times* (April 24, 1954):24.

Manchester, H. "The Prospects for Solar Power." *Reader's Digest* 66:73 (June 1955).

For an in-depth examination of the evolution of photovoltaics, from its nineteenth-century beginnings to the present day, see John Perlin, *From Space to Earth: The Story of Solar Electricity* (Ann Arbor, MI: **aatec publications**, 1999).

# Chapter 1
# Solar Cells: What They Are and How They Work

## WHAT SOLAR CELLS ARE

Solar cells are solid-state devices that absorb light and convert light energy directly into electricity. This is done entirely within their solid structure; solar cells have no moving parts.

### Physical Characteristics

When you look at a solar cell, you see a metallic-blue or black disc covered with thin silvery lines. (Figure 1.1 shows both typical silicon solar cells and the silicon materials from which they are made.) The actual work of converting sunlight into electricity occurs in the dark metallic area because dark colors absorb light more readily. The silvery lines are the **front contact fingers** and are used to make the electrical contact to the front of the cell. The fingers must be very fine so that as little sunlight as possible is blocked out. The back contact is a solid metal layer that both reflects light back up through

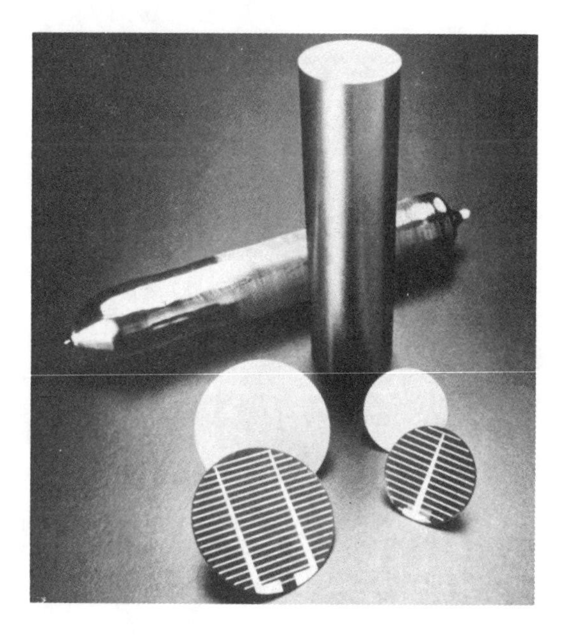

***Figure 1.1.*** Silicon ingot, cylinder, wafers, and finished solar cells. (Jet Propulsion Laboratory, California Institute of Technology)

the cell and makes a good electrical contact. A detailed explanation of how solar cells work appears later in this chapter.

## Electrical Characteristics

In order to use photovoltaic cells, we should have a basic understanding of their electrical characteristics. When illuminated, the solar cell acts somewhat like a battery in that it produces a voltage between front and back. This voltage is developed across a junction that is built into the cell structure. This voltage can be used to produce a current, just like from a battery, but the amount of current is limited by the amount of light falling on the cell.

We can use a simple circuit, as shown in Figure 1.2, to explain the electrical behavior of solar cells. A load resistor connects the front and back of a solar cell. The resistance of the load can be varied from a short-circuit zero resistance to a very high value. Two meters—a voltmeter and an ammeter—measure the voltage developed across the cell and the current passing through the load.

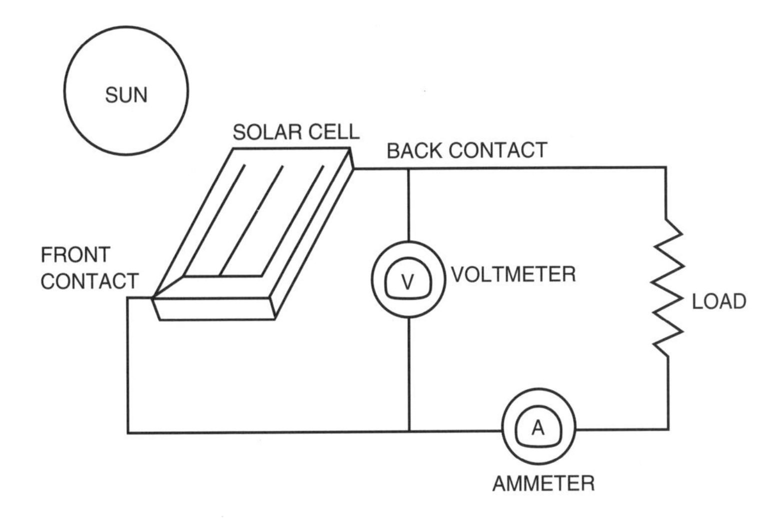


***Figure 1.2.*** A simple circuit to test a solar cell.

If sunlight shines on the cell when the load resistance is very high or the load is disconnected (essentially giving infinite resistance), the voltmeter will read a maximum voltage. This is **open-circuit voltage**. For example, the voltmeter would read 0.58 volts on the typical silicon solar cell. No current is being drawn from the cell under these conditions.

Conversely, if the load resistance is made zero, we will short-circuit the cell (which does a solar cell no harm whatsoever) and draw the maximum possible current from the cell. This **short-circuit current** is directly proportional to the amount of light falling on the cell.

It is possible to adjust the load resistor between these two extremes and measure the voltages and corresponding current produced by the cell under different load conditions. The **current–voltage curve (I–V curve)** in Figure 1.3 shows the results of such an experiment. This figure plots the current on the vertical axis versus the voltage on the horizontal axis. The short-circuit current is shown on the current axis at zero voltage. As the load resistance increases, causing the voltage output of the cell to increase, the current remains relatively constant until the "knee" of the curve is reached. The current then drops off quickly, with only a small increase in voltage,

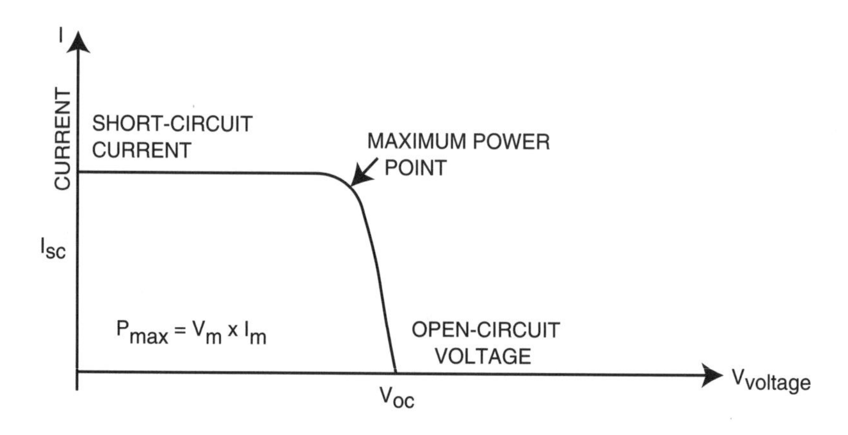


***Figure 1.3.*** Current–voltage (I–V) characteristics of a silicon solar cell.

until the open circuit condition is reached. At this point, the open-circuit voltage is obtained and no current is drawn from the device.

The power output of any electrical device, including a solar cell, is the output voltage times the output current under the same conditions. The open-circuit voltage is a point of no power: the current is zero. Similarly, the short-circuit condition produces no power because the voltage is zero. The maximum power point is the best combination of voltage and current and is shown in Figure 1.3. This is the point at which the load resistance matches the solar cell internal resistance.

Figure 1.4 shows a series of I–V curves for a solar cell under different amounts of sunlight. The peak power current changes proportionally to the amount of sunlight, but the voltage drops only slightly with large changes in the light intensity. Thus, a solar cell system can be designed to extract enough usable power to trickle-charge a storage battery even on a cloudy day.

## Cell Performance Ratings

To compare the performance of different solar cells, the cells are rated at specified amounts and types of sunlight. The most common rating parameter for terrestrial cells is **air mass 1 (AM1)**. This is the amount of sunlight that falls on the surface of the earth at sea level when the sun is shining straight down through a dry, clean atmo-

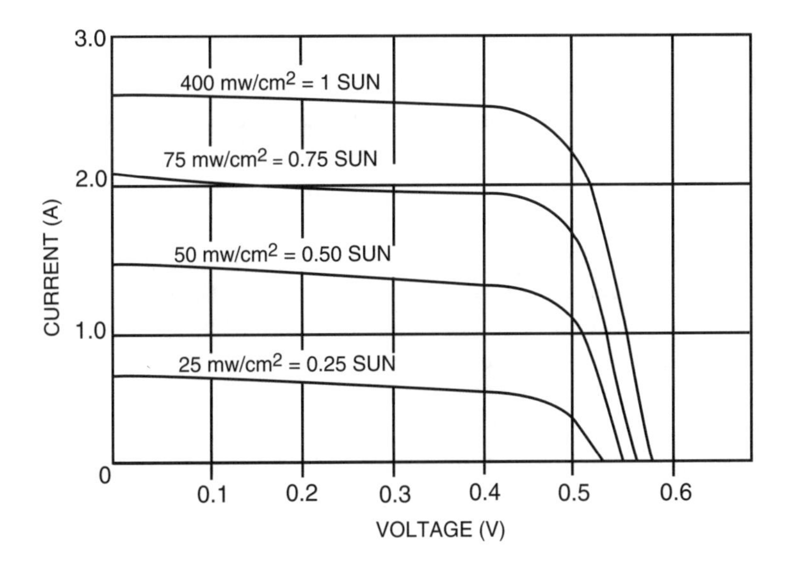


***Figure 1.4.*** Series of I–V curves showing solar cell characteristics versus sunlight intensity.

sphere. The Sahara Desert at high noon, with the sun directly overhead, is a close approximation. The sunlight intensity under these conditions is very close to 1 kilowatt per square meter (1 $kW/m^2$). A closer approximation to the sunlight conditions usually encountered is **air mass 2 (AM2)**, an illumination of 800 $W/m^2$.

## HOW SOLAR CELLS WORK

### The Nature of Sunlight

In order to understand how solar cells work, you first need to know something about the nature of sunlight. All light, including sunlight, is a form of electromagnetic radiation similar to radio waves or microwaves. The sun gives off this radiation simply because it is hot. This **black-body radiation** is composed of a broad mixture of different wavelengths, some of which—the visible spectrum—can be seen by the naked eye and many wavelengths shorter or longer than these. The sun is a black body; if it were cold, it would appear black because it would only absorb radiation.

Figure 1.5 diagrams the solar spectrum. The curve labeled **AM0** illustrates the **air mass 0** spectrum the sun emits as it appears in outer space; the other curve is the **air mass 1** spectrum as seen from the surface of the earth. The difference is caused by our atmosphere. In Figure 1.5, the visible spectrum is in the middle, from 400 to 700 nanometers in wavelength [a nanometer (nm) is $10^{-9}$ meters or one-millionth of a millimeter]. At 700 nm, the visible spectrum appears red; at the shorter wavelength end of 400 nm, it appears violet; the other colors of the rainbow appear between. Our eyes are most sensitive to the wavelengths around 500 nm where the peak of the solar spectrum on earth occurs. Over half of the sun's energy that reaches the earth's surface is in the form of visible radiation.

**Ultraviolet (UV) wavelengths** are shorter than 400 nm and are present in the solar spectrum in small but significant amounts. Ozone and most transparent materials absorb or filter out most of these energetic wavelengths, which is fortunate because short wavelength ultraviolet light can be destructive to organic materials and living things. The small amount of UV light that does make it to the earth's

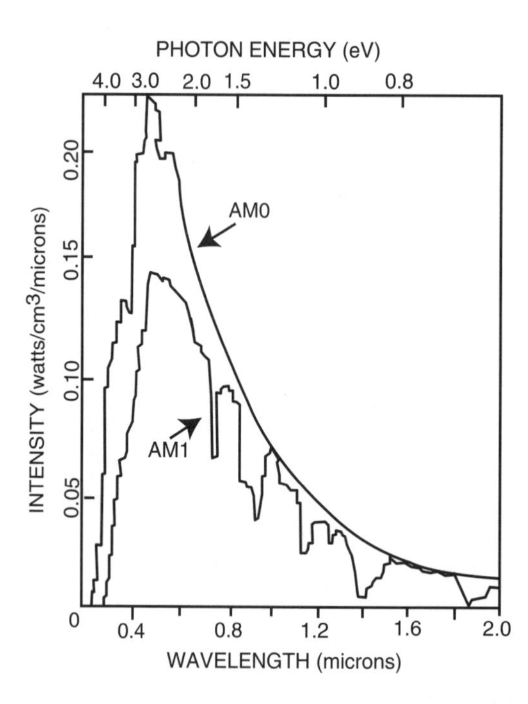


***Figure 1.5.*** The solar spectrum. The dips in the intensity of the AM1 spectrum seen from the surface of the earth are mostly caused by water vapor in our atmosphere.

surface is responsible for the tan—or burn—we receive when exposed to the sun.

**Infrared wavelengths** are longer than 700 nm and, though invisible, are perceived by our skin as radiant heat. As the diagram shows, a great deal of the sun's energy is infrared radiation. Large bands of infrared are absorbed by water vapor, carbon dioxide, and other substances in our atmosphere, but because most of this absorption takes place at longer wavelengths, a solar device that does not effectively collect wavelengths longer than 2 microns (2,000 nm) suffers only a small loss in efficiency.

According to the theory of quantum mechanics, light is composed of particles called photons. Photons, which travel through a vacuum at the speed of light, have no mass, but each is a packet of energy related to the wavelength of the light. The shorter the wavelength, the larger the packet. The energy of individual photons of different wavelengths, as shown at the top of Figure 1.5, is expressed in electron volts. One electron volt is the energy an electron acquires when it accelerates in a vacuum across a potential of one volt. This unit of energy is commonly used by solar cell physicists since it is a convenient size when considering individual electrons.

## Solid-State Physics

Although the solar cell was accidentally discovered in the nineteenth century, and inefficient versions of selenium and cuprous oxide photovoltaic cells were investigated and even used commercially in the early part of this century, it was only after the development of modern solid-state theory and the **band model** of semiconductors that the inner workings of photovoltaic cells have been understood.

The band model of solids, presented here in a simplified version, is based on a crystal with all the atoms fixed in a pattern. The individual atomic nuclei vibrate around a fixed spot in a three-dimensional lattice pattern, but they usually cannot jump completely out of place. Powerful electrostatic forces tie most of the negative electrons in a particular atom very closely to the positively charged nucleus. However, the outermost electrons (called the valence electrons) can be considered "delocalized"; that is, they do not belong to any particular atom, but to the crystal as a whole.

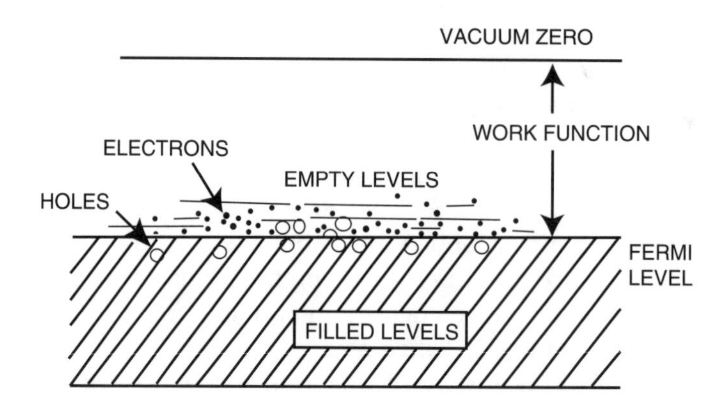


***Figure 1.6.*** Energy level diagram for a metal.

These valence electrons balance the remaining positive charges of all the nuclei. It takes a certain amount of energy to remove one of these electrons from the crystal because of electrostatic attraction. This energy is represented in Figure 1.6 as a vertical distance. In this figure, which illustrates the band diagram for a metal, zero energy is taken as the energy of an electron at rest in empty space. This is called **vacuum zero**. The energies of the electrons in the crystal are all below this zero; the kinetic energy of a fast-moving electron in vacuum would be above the zero.

One conclusion of quantum mechanics is that electrons (or anything else, for that matter) cannot have a continuum of energies but are allowed only certain selected energies (the **energy levels** in the diagram). Another quantum mechanical principle is that no two electrons can occupy the same energy level.

It is possible to calculate the number and positions of the energy levels of a particular metal; if a count is made, it turns out that each atom contributes more energy levels than electrons. (This also is shown in Figure 1.6.) Since the band of valence levels is greater than the number of electrons, some levels must be empty. The electrons would preferentially occupy the lowest lying levels. At absolute zero, when none of the electrons are moving, they would fill the band to the fermi level, where the probability of finding an electron is one-half. In order to move, the electrons must have a bit more energy (their kinetic energy) and would occupy a level just above the **fermi level**. Since there are many empty levels at this energy, the electron has a large number of possible kinetic energies and can move very freely throughout the crystal. These moving electrons can produce an electric current, so a metal would be expected to be a good conductor of electricity and capable of carrying a large current.

Each electron from below the fermi level that jumps to a higher empty level leaves behind an empty level. These empty levels in the "sea" of electrons are called **holes**. A hole is a level that an electron could occupy but currently does not. A hole "moves" when an electron falls into it, creating a new hole. Thus, a hole can be seen as a particle similar to an electron with a positive charge, a mobility, and even an effective mass. Holes can carry electric current and in some metals are the dominant carriers.

If the number of available levels is exactly the same as the number of electrons for each atom, the situation depicted in Figure 1.7 results. This figure also shows a band of completely empty energy levels at some higher energy. Since the number of excited electronic states in any atom is theoretically infinite, there will always be such a **conduction band**, sometimes separated from the **valence band** by a **band gap**. Because the valence band is completely filled, there are no nearby energy levels for an electron to occupy. The electrons

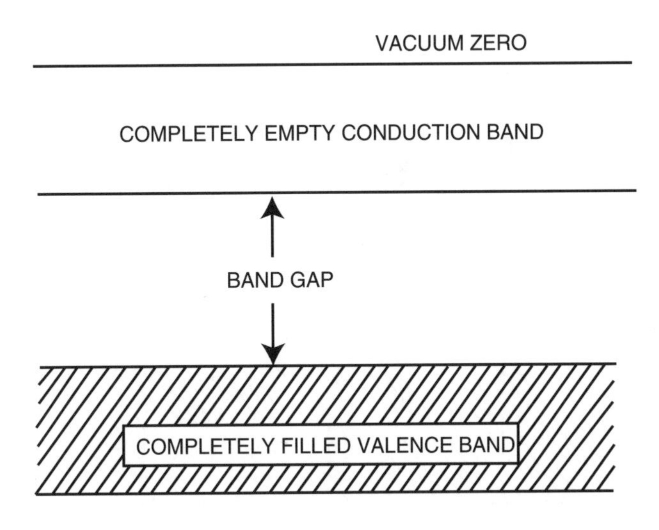


***Figure 1.7.*** Energy level diagram for a perfect insulation.

cannot move and therefore cannot create any holes. Thus, no conduction can take place and the material is called an insulator. The energy of the band gap is in electron volts, as are all the energies shown in the diagrams.

If the band gap is sufficiently narrow (below 2.5 electron volts or so), it is possible for a particularly energetic electron to jump from the top of the valence band to one of the empty levels in the conduction band (see Figure 1.8). The electron might be thermally excited by the vibrations of the atoms in the crystal lattice or a photon of light of sufficient energy might be absorbed by the crystal and cause the electron to jump. Once the electron is in the upper conduction band, it is free to move and can act as a carrier of electricity. In addition, the hole left behind in the valence band can also be a current carrier. A material that has these properties is called a semiconductor.

In Figure 1.8, the fermi level is in the middle of the band gap. Even though there are no energy levels in the middle of the band gap, one can have the concept of a fermi level there. Were there an energy level at this place, it would be occupied by an electron half

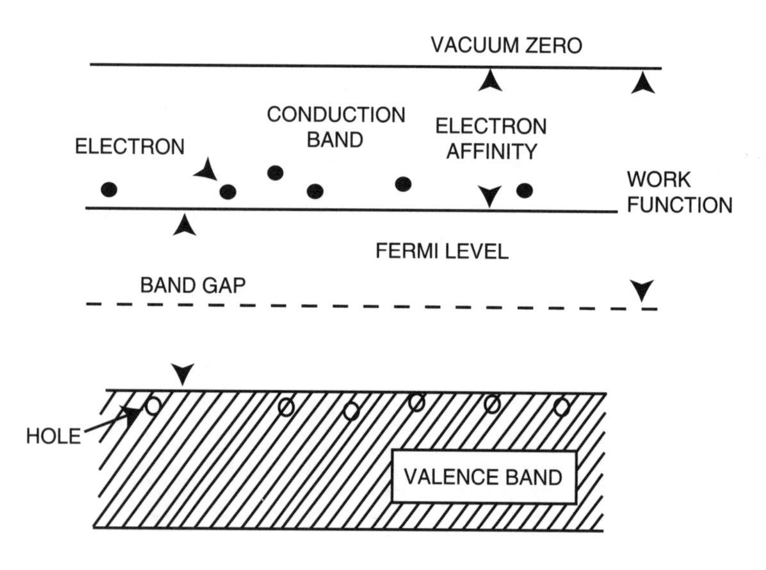


***Figure 1.8.*** Energy band diagram for an intrinsic semiconductor.

the time, on average. Above the fermi level, fewer levels are occupied; below it, most are. The unoccupied levels are holes.

As the temperature of a solid increases, the most energetic electrons and holes can be found farther from the fermi level, but the position of the fermi level remains fixed. Only the most energetic electrons and holes are actually occupying levels. (The most important point to remember is that the closer a band is to the fermi level, the more conduction electrons—or holes, if *below* the fermi level—that will be found in the band.) For a more complete mathematical discussion of these solid-state concepts, see Kittel's *Introduction to Solid-State Physics*, 5th Edition.

## Doping

The electrical conductivity discussed in the last section is called **intrinsic conductivity** because it is an intrinsic property of the particular material and its crystal lattice. The ideal **intrinsic semiconductor** cannot exist in reality because all materials have some impurities, no matter how carefully prepared. The impurity atoms will occupy various spots in the crystal lattice and disturb its perfection. In addition, these foreign atoms usually contribute a different number of levels and/or electrons to the system than the normal atoms. For example, in a single crystal of silicon each silicon atom contributes exactly 4 valence levels, 4 conduction levels, and 4 electrons since the valence of silicon is 4.

If an atom of boron (which has a valence of 3) is added to the crystal, it will contribute only 3 electrons to the system. It will also contribute 3 valence levels, plus 1 level which will be created in the band gap slightly above the top of the valence band. This situation is diagrammed in Figure 1.9 (left). These extra levels, called **acceptor levels**, are capable of accepting an electron from the valence band. Since the acceptor level is associated with a particular impurity atom, the electron that occupies it is trapped and cannot move. However, the hole created when the electron leaves the valence band can move and carry an electric current. Silicon doped with boron makes a **p-type semiconductor**, since **p**ositive holes carry the current.

The process of deliberately adding a known impurity to a pure semiconductor is called **doping** and the resulting material is an **extrinsic semiconductor**. Similarly, if a small amount of phosphorus

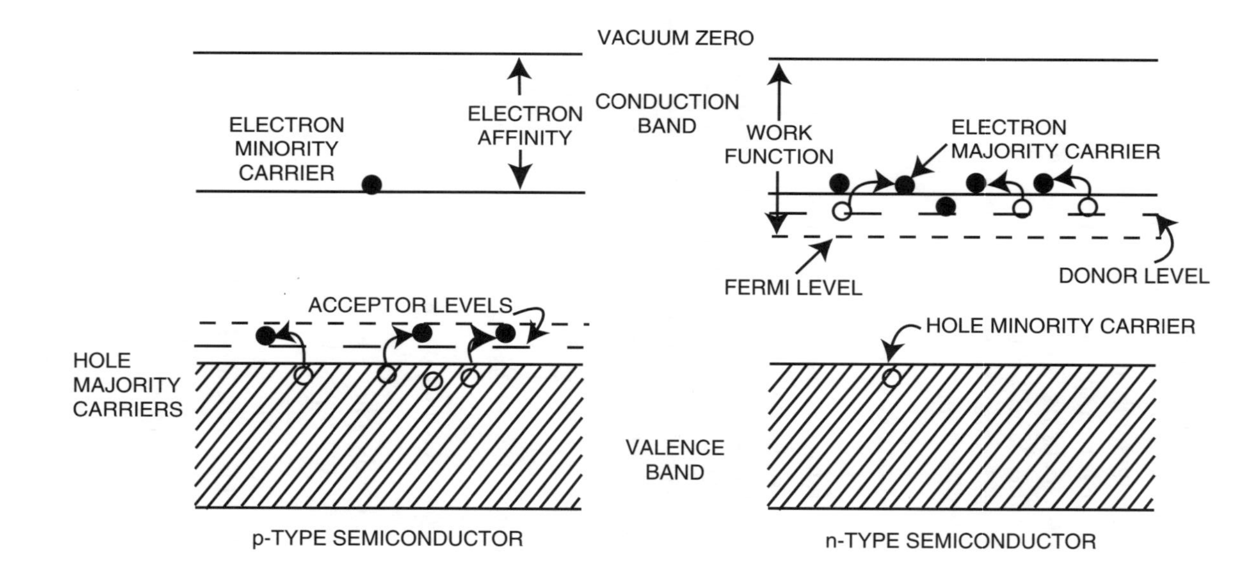


***Figure 1.9.*** Energy band diagram for extrinsic (doped) semiconductors.
*Left:* p-type semiconductor. *Right:* n-type semiconductor.

is added to the silicon crystal, the atoms of phosphorus (which have 5 valence electrons) will contribute 4 electrons and 4 levels to the valence band, and the extra electron will occupy a level near the bottom of the conduction band as shown in Figure 1.9 (right). It takes very little energy for the extra electron to jump to the conduction band from this **donor level**, so-called because it can donate conduction electrons to the system. This type of material is an **n-type semiconductor**, since **n**egative electrons carry the electric current.

Notice that Figure 1.9 (left) shows p-type material: the fermi level is located near the valence band and the vast majority of current carriers are holes. Figure 1.9 (right) shows n-type material: the fermi level is near the bottom of the conduction band and the majority of carriers are electrons. In p-type semiconductors, the positive holes are the **majority carriers** and the electrons are the **minority carriers**. In n-type semiconductors, the negative electrons are the majority carriers and the holes are naturally called minority carriers.

Finally, note the quantities marked **work function** and **electron affinity**. The work function is the energy difference between the fermi level and vacuum zero; the electron affinity is the energy difference between the bottom of the conduction band and vacuum zero. These definitions hold true for all semiconductors. For metals, the work function and electron affinity have the same value.

## JUNCTION PHOTOVOLTAIC CELLS

### Homojunctions

Figure 1.10 shows the construction of a typical n-on-p junction solar cell. This is the most common type of cell and is made by taking a wafer of p-type single-crystal silicon and diffusing phosphorus atoms into the top surface. This can be done by heating the wafers in a diffusion furnace in the presence of phosphorus-containing gas. The phosphorus atoms have a valence of 5. This means that each phosphorus atom has 5 electrons in its outermost region compared to silicon atoms which have 4. The extra electron is very loosely held near the atom, producing donor levels in the top layer of silicon, making it n-type. The n-type silicon on top of the p-type silicon produces a junction where the two types come in contact. The back

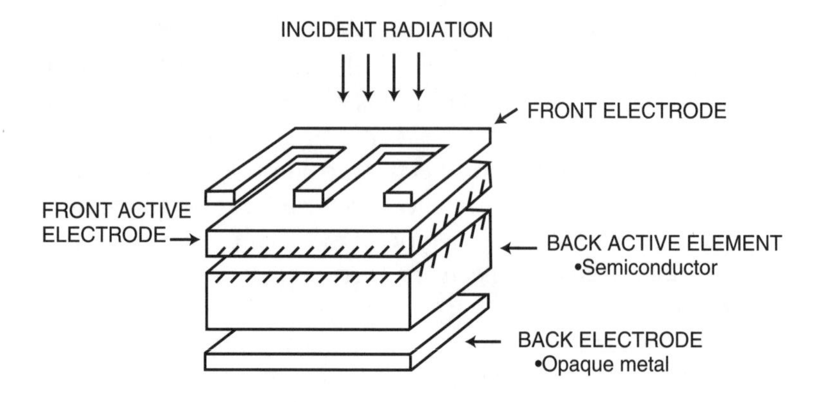


***Figure 1.10.*** Cross section of a solar cell showing its construction.

contact and the top fingers shown are used to connect the solar cell to the external circuit. A more detailed description of the construction of these cells is given in the next chapter.

The energy band diagram for this n-on-p cell (called a **homojunction** solar cell) is shown in Figure 1.11. This diagram is made by overlaying the diagram for the n-type semiconductor with the diagram for the p-type material and then aligning the fermi levels. Notice that now there is a difference in the vacuum zero levels for the two materials. This represents the **built-in potential** that exists between the materials because of the difference in their work functions. Between the n- and p-layers is a **depletion layer** where there are essentially no carriers—neither electrons nor holes. This layer, where the bands are bent, is the actual junction. The front and back metal contacts must be **ohmic contacts**: that is, contacts that do not impede the flow of electrons into or out of the semiconductor. In the dark, if the junction device is connected to a battery so the front fingers are made positive and the back negative, the situation shown in Figure 1.12 (top) would result. The applied potential is added to the built-in potential and the uphill barrier for the electrons is increased; this results in no current flow except for the tiny current of electrons thermally excited from the valence to the conduction band. At room temperature this current would be a fraction of a microamp. On the other hand, if the battery is reversed so that the front is negative and the back positive, the applied potential can be made large enough to cancel out the built-in potential. As Figure 1.12 (bottom)

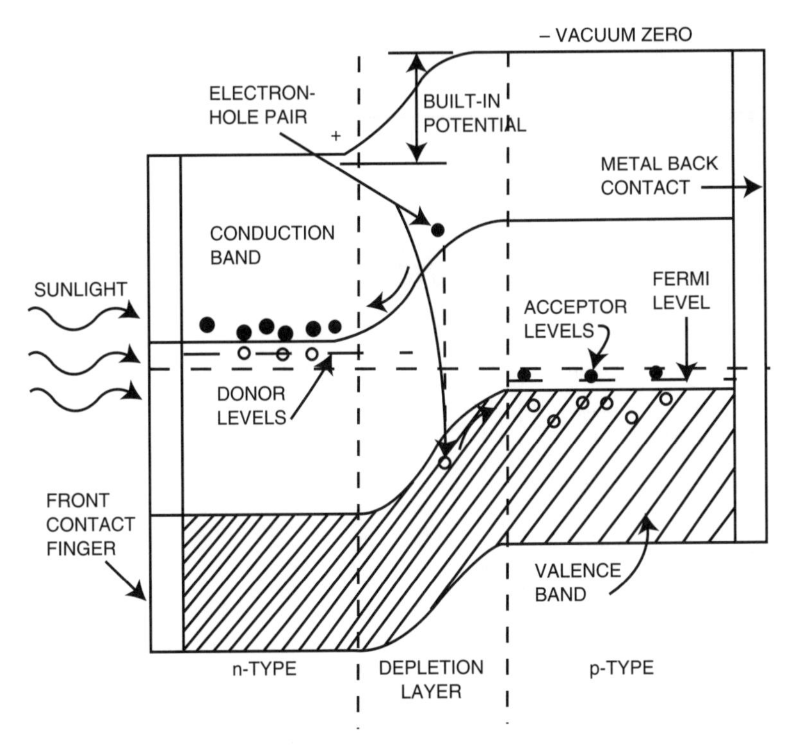


***Figure 1.11.*** Band diagram for p–n homojunction solar cell.

shows, there is no longer any barrier to the flow of either holes or electrons and the device becomes a good conductor of electricity. So, a solar cell is a cell is a **junction diode** or **rectifier**, letting electricity pass in only one direction in the dark. The "dark" curve in Figure 1.13 is a typical current voltage plot for this type of junction.

If light falls on a solar cell, those photons with energy greater than the band gap width can excite an electron from the valence to the conduction band and create a hole-electron pair. The conduction electron, if created in the depletion layer, will slide "downhill" toward the front n-layer and the hole will slide "uphill" (like an air bubble in water) to the p-layer. Thus, the charge carriers will be separated and current will flow. The ohmic front contact allows the electron to flow through an external circuit to the back contact where it can recombine with a hole, leaving everything as it was. Even if the conduction electron is created away from the junction, if it happens

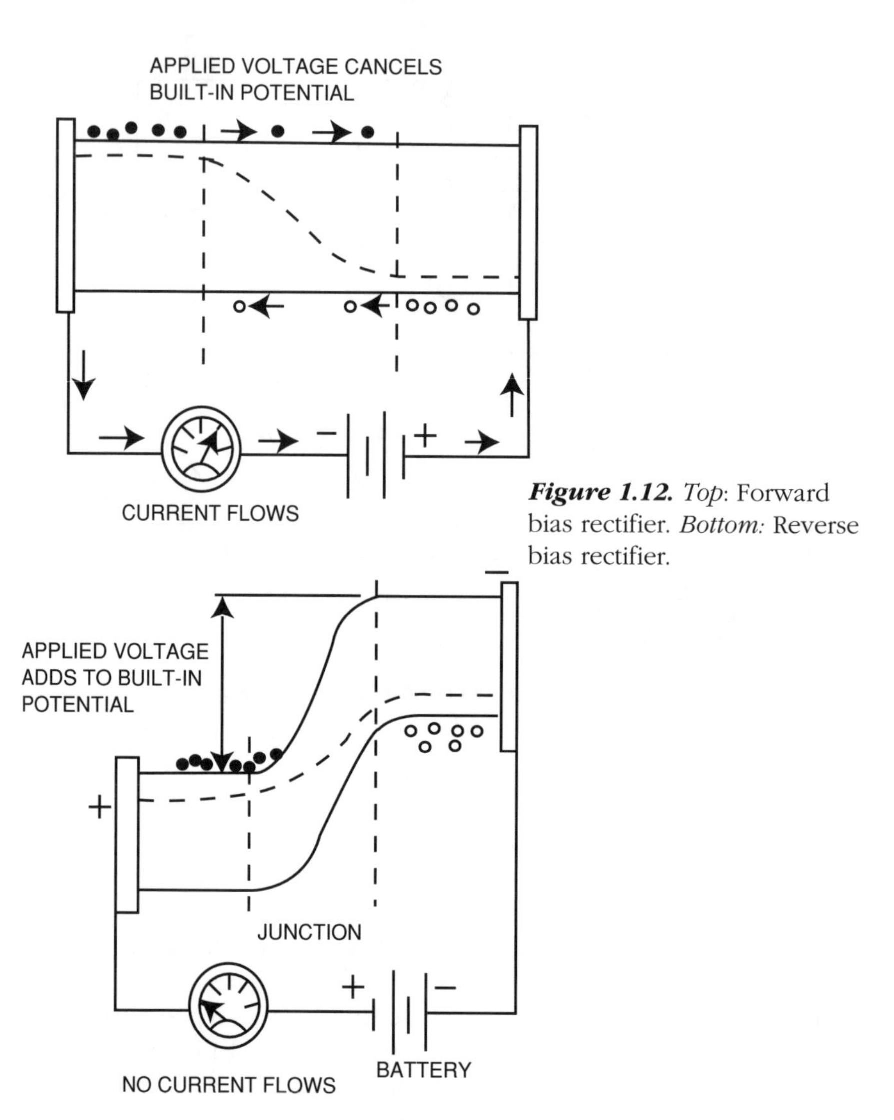


***Figure 1.12.*** *Top*: Forward bias rectifier. *Bottom:* Reverse bias rectifier.

to drift toward the junction it will be carried down as before and produce a current in the external circuit.

If the external circuit is simply a wire and has no appreciable resistance, the current that flows is the short-circuit current ($I_{SC}$) and is directly related to the number of photons of light being absorbed

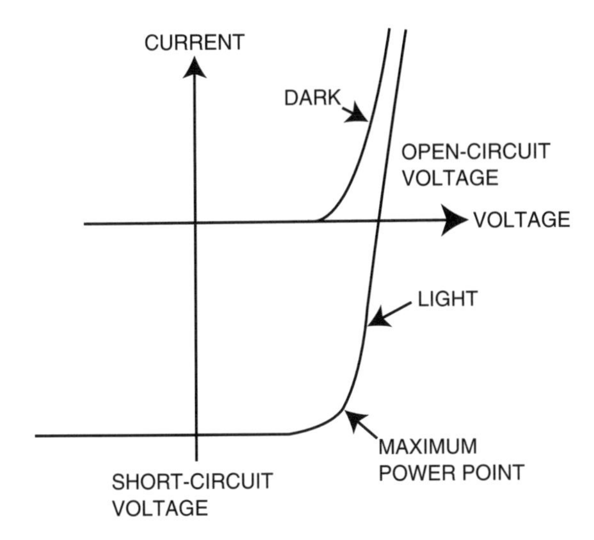


***Figure 1.13.*** Current–voltage plots for a solar cell both in the dark and illuminated.

by the cell. This is why short-circuit current is directly proportional to light intensity.

If there is no external circuit, the incoming photons will still create hole–electron pairs and they will still travel downhill to the p- and n-layers, but they will pile up there because there is no external wire to complete the circuit. The extra negative electrons in the n-layer and the extra positive holes in the p-layer will cause a situation analogous to that shown in Figure 1.12 (bottom), except that now the battery is the solar cell itself. Soon the voltage difference between the front and the back of the cell will become large enough to flatten the barrier sufficiently to allow some holes and electrons to leak back. When the number of carriers leaking back is equal to the number being generated by the incoming light, an equilibrium voltage has been reached. This is called the **open-circuit voltage ($V_{oc}$)** and is shown on the light current–voltage curve on Figure 1.13. The open-circuit voltage varies only slightly with large changes in light intensity.

If the resistance of the external circuit can be varied from zero (short circuit) to extremely high (open circuit), the entire current voltage curve can be plotted. As stated at the beginning of the chapter, the power emitted by the solar cell is defined as the voltage times the current:

$$P_{(watts)} = V_{(volts)} \times I_{(amps)}$$

For the short circuit, there is no voltage and hence no power output from the cell. Similarly, for the open circuit, there is maximum voltage but no current, so again there is no power output. The best combination of voltage and current to produce the maximum power lies somewhere between. That point is shown at the bend in Figure 1.13.

One way to judge the quality of a solar cell is by measuring the **fill factor (ff)**. The fill factor is the actual maximum power divided by the hypothetical "power" obtained by multiplying the open-circuit voltage ($V_{OC}$) by the short-circuit current ($I_{SC}$)

$$ff = \frac{P_{max}}{V_{OC} \times I_{SC}}$$

A good silicon solar cell will normally have a fill factor above 0.75.

## FACTORS WHICH INFLUENCE CELL EFFICIENCY

### Band Gap Width

One of the most important factors influencing theoretical efficiency is the band gap width of the semiconductor material. Remember that only photons of energy greater than the band gap can produce current carriers. This means that only light of wavelengths shorter than a critical maximum can be used to produce a photocurrent. In fact, most semiconductors are transparent to light that is more red than this cutoff wavelength; photons will pass right through the material without interaction. Therefore, small band gap materials will absorb more of the red end of the solar spectrum and would appear to be more efficient users of sunlight. However, the open-circuit voltage of a cell is limited by the band gap width of the cell material. Also, the energetic photons of short wavelength cannot produce any more current than the photons barely able to raise an electron to the conduction band, thus they are used less efficiently by narrow band gap materials (the excess energy of the photon shows up as heat and it heats the cells). The best cells will have compromise band gap widths.

In 1953, Dr. Dan Trivich of Wayne State University made the first theoretical calculations of the efficiencies of various materials of different band gap widths based on the spectrum of our sun. Figure 1.14 is an updated version of those calculations. Although the diagram shows that silicon has too narrow a band gap for optimum efficiency, more than 35 years of development have pushed the actual working efficiency closer to the theoretical.

Care must be taken to distinguish between quantum efficiency, the number of current electrons created and delivered to the front contacts compared to the number of photons falling on the solar cell, and energy efficiency, the total electrical power output compared to the input power flux of light energy. For example, a good crystalline silicon cell may have a quantum efficiency of 85% but a total energy efficiency of 19% because the high-energy blue-light photons only produce the same energy in electrons as do the infrared photons.

## Recombination

A major factor that influences attainable efficiency is recombination. Not all the hole-electron pairs created make it all the way to the

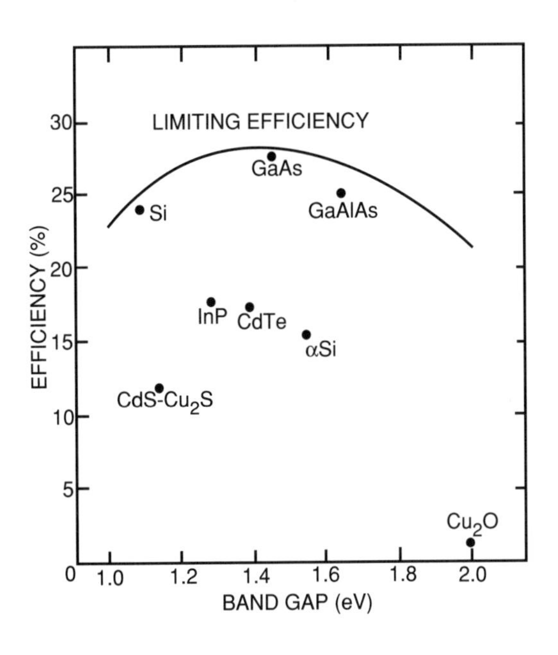


***Figure 1.14.*** Efficiencies versus band gap for various semiconductor solar cell materials. Dots are one-sun measured values.

metal contacts; some of the carriers will recombine with a carrier of opposite sign, thus eliminating both. This usually happens at a defect or impurity in the crystal called a **recombination center**. It is especially common to find these centers at the junctions or at the surface of the material, but recombination can happen even in the main body of the material. (These are called **junction**, **surface**, or **bulk recombination**, respectively.) Using only high-quality single-crystal wafers to make solar cells will minimize recombination. Many manufacturers have learned how to suppress the recombination centers formed at crystalline boundaries, leading to the commercial use of polycrystalline silicon wafers in high-quality solar cell production.

### Reflectivity and Light Absorption

Most semiconductor materials have a high reflectivity of the light they should be absorbing. To decrease reflection loss, most manufacturers etch the surface of the cell to roughen or "texturize" it and apply antireflection coatings. These coatings, similar to those on good camera lenses, are usually silicon monoxide or silicon nitride for silicon cells.

Light is also reflected by the shiny metal fingers used as the top contacts on the solar cells. The design of the finger pattern is a compromise between a very open pattern that will not block out much light and one of very low resistance that wastes little power output resistance-heating the fingers.

## HETEROJUNCTIONS

The n- and p-layers in a solar cell do not have to be made of the same material. Although **heterojunctions** are composed of two different semiconductors (see Figure 1.15), they operate in essentially the same way as homojunctions. The two different semiconductors offer a great deal of design flexibility; if the top material has a wide band gap, it can serve as a window to let the light penetrate to the second layer. The window layer is heavily doped to make it a good electrical conductor, sometimes eliminating the need for metallic front fingers. The second layer is selected for its efficiency as an absorber of sunlight. With the proper choice of materials, the device can have a larger built-in potential and, hence, a higher open-circuit voltage

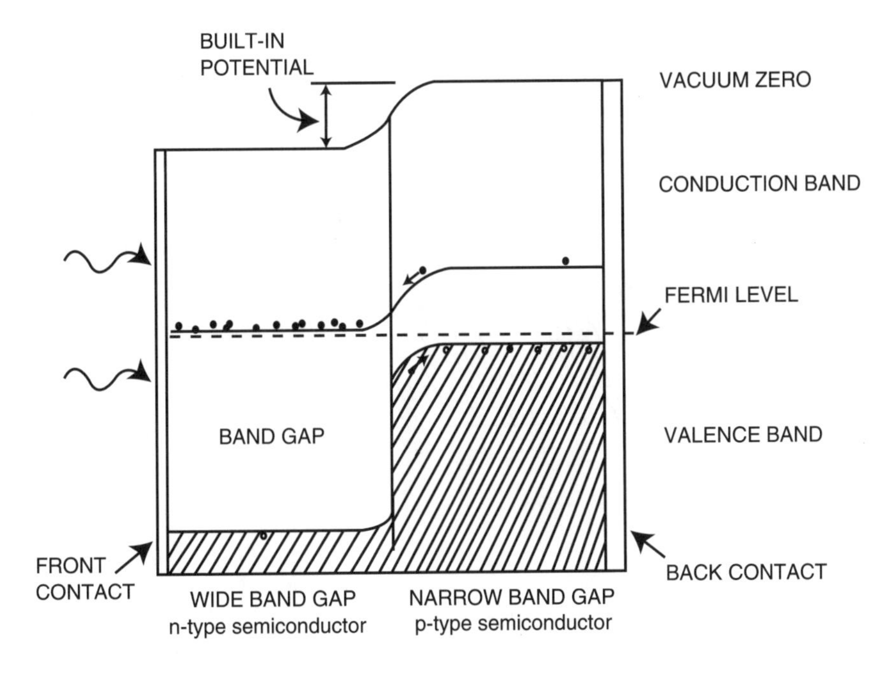


***Figure 1.15.*** Band diagram for a heterojunction solar cell.

than a homojunction made from the narrow band gap semiconductor. Also, some semiconductors, such as cuprous oxide or tin oxide, are available in only one form—either p- or n-type—and cannot be made into homojunction cells. Heterojunctions can exploit the best properties of such materials. Examples of heterojunction solar cells include a cadmium sulfide/cuprous sulfide system and the three-layer cadmium sulfide/cadmium telluride/zinc telluride cell. In the latter, the window is n-type CdS, the light absorber is a thin layer of nearly intrinsic CdTe, and the p-layer is highly conductive ZnTe. Chapter 6 discusses new developments in heterojunction systems.

## SCHOTTKY BARRIER JUNCTIONS

The metal-to-semiconductor or Schottky barrier junction (Figure 1.16) is formed when a low work-function metal is placed on a p-type semiconductor. A high work-function metal on an n-type semiconductor will also produce a Schottky barrier, one with a band diagram resembling an upside-down version of the one shown. The impor-

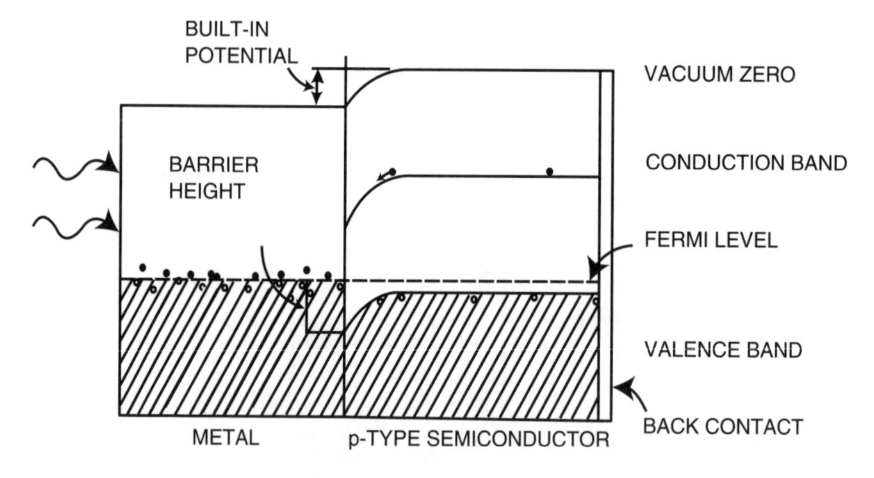


***Figure 1.16.*** Band diagram for a Schottky barrier junction solar cell.

tant point is the creation of a barrier to the flow of majority carriers from the semiconductor to the metal and to the flow of all carriers from the metal to the semiconductor. Also, the band bending in the semiconductor produces a built-in potential that determines the open-circuit voltage in the solar cell. The built-in potential is simply the difference in work function between the metal and the semiconductor while the barrier height is the energy difference between the work function of the metal and the top of the valence band (for p-type semiconductors) or the bottom of the conduction band (for n-type). The quality of the solar cell, as measured by the reverse dark current and the fill factor, is dependent in part on the barrier height. If a high work-function metal is put onto a p-type semiconductor (or a low work-function metal onto an n-type material), an ohmic contact with no barrier at all is the result.

The first solar cells—the cuprous oxide cell and the selenium cell—were both Schottky barrier cells. However, an efficient and inexpensive Schottky barrier solar cell has yet to be constructed.

## ADVANCED SEMICONDUCTOR DEVICES

There are a number of other semiconductor junctions that have been considered for use in solar cells. Two promising devices are the metal–

insulator–semiconductor (MIS) and semiconductor–insulator–semiconductor (SIS) junctions where a thin insulating layer is placed in the middle of the junction. Current carriers pass through this insulator by a quantum mechanical process called **tunneling**. Some scientists believe that all Schottky barrier junctions have an insulating layer and are MIS structures. If the insulating layer is of the proper thickness, around 20 Å, the open-circuit voltage of a cell can be increased without any significant loss in short-circuit current. For instance, MIS structures using cuprous oxide, aluminum oxide, and aluminum showed a brief open-circuit voltage of over one volt, but the structure was unstable and no further work on this combination has been attempted.

## RECOMMENDED READINGS

Angrist, Stanley W. "Photovoltaic Generators." In *Direct Energy Conversion*, 2nd ed., Chapter 5 (Boston: Allyn & Bacon Inc., 1971).

Backus, Charles E., Ed. *Solar Cells* (New York: Institute of Electrical and Electronics Engineers, 1976).

Commoner, Barry. "The Solar Transition." *The New Yorker* 55:53–54, Part I (April 23, 1979); 55:46–48, Part II (April 30, 1979).

Chalmers, Bruce. "The Photovoltaic Generation of Electricity." *Scientific American* 235:34–43 (October 1976).

Hovel, Harold. "Solar Cells." In *Semiconductors and Semimetals*, Volume 11, R. K. Willardson and A. C. Beer, Eds. (New York: Academic Press, Inc., 1975).

Kittel, Charles. *Introduction to Solid-State Physics,* 5th ed. (New York: John Wiley & Sons, 1976).

Landsberg, P. T. "An Introduction to the Theory of Photovoltaic Cells." *Solid-State Electronics* 18:1043–52 (1975).

*Photovoltaics Technical Information Guide*, 2nd ed. (Golden CO: NERL, NTIS, U.S. Department of Commerce, 1988)

Zweibel, Kenneth. *Harnessing Solar Power: The Photovoltaics Challenge* (New York: Plenum Press, 1990).

## Chapter 2
# How Solar Cells Are Made

The present processes for manufacturing solar cells are complicated, energy-consuming and expensive, although considerable progress has been made in the last decade to lower the cost and energy consumed per cell. This chapter describes the current techniques for **single-crystal**, **polycrystalline**, and **amorphous** photovoltaic cell manufacture and examines improvements now under development. The step-by-step procedure for making cells follows.

### SILICON

You start with sand. Silicon, the second most abundant element in the earth's crust, is present in almost all rocks and minerals, but the most convenient source is silicon dioxide in the form of white quartzite sand. If the sand is pure enough, the subsequent (and costly) purification steps are less complicated.

## Metallurgical-Grade Silicon

Sand is reduced to silicon in an electric arc furnace. A carbon arc reacts with the oxygen in the silicon dioxide to form carbon dioxide and molten silicon. This common industrial process produces a metallurgical-grade silicon with about 1% impurities that has a number of uses in the steel and other industries. Though relatively inexpensive, this silicon contains far too many impurities for use in the microelectronics industry. Many impurities are the result of using contaminated sand and could be eliminated by the careful choice of raw materials; others are introduced during processing and could be eliminated by quality control.

## Semiconductor-Grade Silicon

The electronic properties of silicon semiconductor devices are determined by a small percentage of **dopant** atoms in the crystal lattice. In order for the devices to be built, the number of impurities in the silicon starting material must be few in comparison to the dopants to be added. This means that semiconductor-grade silicon must be hyperpure (the residual impurities are measured in parts per billion).

The most common way to produce silicon of very high quality is by the thermal decomposition of silane or some other gaseous silicon compound. A seed rod of ultrapure silicon is heated red-hot in a sealed chamber and the purified silicon compound is admitted. When the molecules of this compound strike the rod, they break down to form elemental silicon which then builds up on the rod. When the rod has grown to the desired thickness, it is removed; the rod is now ready to be made into a single crystal.

An alternative process for purifying silicon is **zone refining**. In this process, a rod of metallurgical-grade silicon is clamped in a machine with a moving induction heater coil. The coil first melts a zone through the entire rod near the bottom and then slowly travels upward. The silicon then melts at the top edge of the zone and solidifies at the bottom edge. When the liquid solidifies, impurities in the melt are excluded from the new crystal lattice so that the molten zone picks up the impurities and sweeps them to the top of the rod, which is then discarded.

***Figure 2.1.*** Semiconductor-grade silicon. This highly refined material is purer than fine gold and is becoming more expensive as supplies fail to keep up with demand. (Siemens Solar Industries, Camarillo, CA)

Both methods of purifying silicon are expensive and energy-consuming. The resulting semiconductor-grade silicon presently costs $25 per kilogram; this is down from the $80 per kilogram cost fifteen years ago. Single-crystal solar cells are currently made from this type of silicon, but polycrystalline and amorphous silicon cells use lesser-grade materials. The latter part of this chapter describes these manufacturing processes.

## GROWING SINGLE CRYSTALS

The boundaries between crystals in silicon act as traps for the electrical current carriers. To avoid these traps and to ensure the best performance, all semiconductor devices and some solar cells are made from single crystals of silicon. The techniques for growing silicon crystals have developed to the point that crystals six inches in diameter and six feet long are routinely grown by the **Czochralski process**. In this method, a seed crystal is dipped into a crucible of molten

***Figure 2.2.*** A single crystal of pure silicon being pulled from a crucible of molten material. Virtually all solar cells commercially produced are made by the Czochralski process. (Siemens Solar Industries, Camarillo, CA)

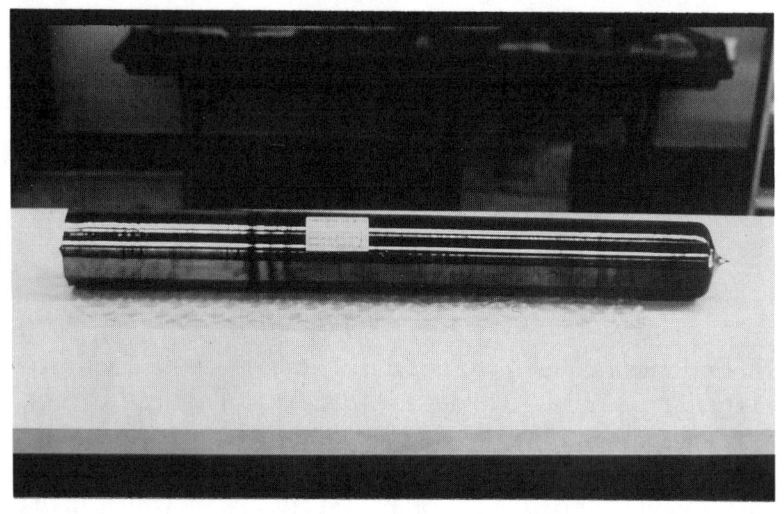

***Figure 2.3.*** An ingot produced by the Czochralski process. The standard diameter of ingots has increased from 2.25 and 3 inches to 4.25 and even 6 inches as the technology has improved. (Siemens Solar Industries, Camarillo, CA)

silicon and slowly withdrawn, pulling a large round crystal as the molten material solidifies on the bottom of the seed. Feedback controls adjust the pulling speed and the temperature of the melt to produce a crystal of a given size. The total mass of the crystal is determined by the size of the pulling machine and the amount of material the crucible can hold. A small amount of dopant is usually added to the silicon during this step to produce the desired electronic properties.

## SLICING THE CRYSTAL INTO WAFERS

Because most semiconductor devices, including photovoltaic cells, must be extremely thin, the silicon crystal has to be sliced into thin wafers. The thickness of the wafer—on the order of 300 microns (about 0.012 inch)—is determined chiefly by the fragility of the material; thinner wafers would be too difficult to handle. Slicing or "wafering" is usually done with a single thin-metal blade coated with diamonds. On most of these saws, the cutting edge is the inside edge of a washer-shaped blade. The outer edge of the disc of these "i.d." saws is rigidly supported in a heavy ring to reduce flexing and to give a straight, smooth cut. A liquid lubricant is pumped continuously into the saw cut to cool the materials and wash away sawdust. Although the process is automated, it is time-consuming: one crystal will yield thousands of wafers—each cut one at a time. It is also wasteful because half of the high-purity silicon is lost as sawdust. Some manufacturers, notably Solarex (now part of BP Solar), have developed cutting techniques that use multiple wires and a diamond slurry cutting fluid which yield thinner wafers and less waste.

## POLISHING AND ETCHING THE WAFER

The surfaces of a solar cell must be specially prepared to ensure the proper electrical and optical properties. Originally, solar cells were highly polished by using fine abrasives and chemical polishing etchants. A number of these mixtures, first developed for the microelectronics industry, leave a mirror-like surface that forms a good junction with the metals used for electrical contacts. Texturized surfaces, which

***Figure 2.4.*** Slicing a silicon single crystal with a diamond i.d. saw. This process is time-consuming and wasteful: the saw cuts are as wide as the wafers produced. (Siemens Solar Industries, Camarillo, CA)

look like mountain ranges under the high magnification of an electron microscope, have recently been found to be more effective absorbers of light. Some manufacturers, such as Kyocera, have dispensed with grinding and polishing completely, relying on a simple etching procedure to produce the desired surface. Other surface treatments, such as ion milling, are being examined by manufacturers.

## FORMING THE p–n JUNCTION

The starting material for most photovoltaic cells is p-type silicon. During crystal growth, a small amount of boron is incorporated into the crystal lattice. To make a p–n homojunction cell, the top few microns of the wafer must be made n-type. This is achieved by incor-

***Figure 2.5.*** A diffusion furnace in which the p–n junction is formed. One furnace can process over 1,000 cells at a time. (Motorola Semiconductor Group, Phoenix, Arizona)

porating more phosphorus atoms into the top layer of material than boron atoms. The excess phosphorus, being an electron donor, makes the layer n-type. To do this, a rack of wafers is heated to a high temperature in a **diffusion furnace** in the presence of a phosphorus-containing gas. At a high temperature, but one that is still well below the melting point of silicon, the individual atoms will move and vibrate wildly within the crystal lattice and foreign atoms striking the surface will diffuse slowly into the bulk of the material. If the temperature and time of the exposure to the gas are properly controlled, a uniform junction can be formed a known distance into the wafer.

The surface that is to be the back of the cell must be protected to keep a junction from forming there also. Something as simple as sealing two wafers back to back can accomplish this. Sometimes a p+-p junction is diffused into the back of the cell to create a back surface field and improve the collection efficiency for the light-generated electrons.

A newer method of forming the front junction is **ion implantation**. This method has been used to make integrated circuits for a number of years, but has been applied to solar cells only recently. The ion implanter, a machine similar to a small particle accelerator, shoots individual ions at the surface of the wafer. (In this case the ions are atoms with one electron missing.) The depth to which the ions penetrate the front of the silicon can be controlled by changing the speed at which the ions hit the surface, so that the thickness and characteristics of the top layer of the cell can be carefully tailored. As automated ion implanters become more common, this junction formation technique will become an important one, especially for high-quality cells.

## APPLYING THE FINGERS AND BACK CONTACT

Once the junction is formed, you have a working solar cell. To make the cell usable, however, contacts must be placed on the wafer so that the device can be connected to an external circuit. Both the front and back contacts must be ohmic contacts with as low a resistance to electric current as possible. The contact material must also adhere very well to the silicon and must be able to withstand soldering (or, in some cases, spot-welding). The back contact can be a solid metal coating since light does not have to pass through this area. In fact, a shiny metal contact will reflect back through the silicon any light that is not absorbed in the first pass through the cell; this allows a thinner wafer of silicon to be used.

A great deal of work has been done in contact design and fabrication. Each manufacturer has its own technique for producing inexpensive, reliable, rugged contacts. The most reliable seem to be those made from palladium/silver layers that are vacuum-evaporated through a mask or photoresist to make the initial thin layer. Once the pattern is established, a thicker metal layer is electroplated on top. Fingers made in this way can be very narrow and close together so that they block little light while maintaining a small resistance.

Another contact system incorporates electrodeless nickel plating. The solar cell is covered with wax using a silkscreen that exposes the silicon where the fingers are to be placed. Then the cell is dunked into a bath containing nickel compounds. A chemical reaction deposits pure nickel on the uncovered silicon, producing the fingers. After the wax is removed, the fingers are tinned with solder. While simple and relatively inexpensive, the contacts this method produces do not always adhere to the silicon as well as would be desired.

Other methods of applying contacts include silkscreening a metal-containing paint onto the wafer and then baking to drive off the organic components which leaves a metal frit. Although present day cells have much finer fingers and much better tab adherence than solar cells made only a few years ago, extensive research continues to be devoted to improving contact systems.

## THE ANTIREFLECTION COATING

Silicon is a very shiny material with a grey metallic appearance. It reflects 35% or so of the light that falls on it. To the solar cell, this is lost light that could have generated electricity. To lessen this waste, all solar cells are coated with an antireflection coating, the same type of coating that has been used for decades on the lenses of good cameras, binoculars, and other high-precision optics. These coatings are transparent layers with a thickness of about one-fourth the wavelength of light; a typical thickness is less than 100 nanometers or 0.1 micron. The coating could be made of any hard transparent material but the preferred optical coatings adhere well to silicon, have the right **refractive index** (they bend light properly), and are relatively easy to apply in accurately controlled layers.

Silicon monoxide, titanium dioxide, and some other optical coatings can be applied to the silicon in a vacuum coating process, such as **vacuum evaporation** or **sputtering**. In vacuum evaporation, a substance is heated to a high temperature in a vacuum system. The molecules of the substance "boil" off the surface and travel in straight lines until they encounter a cool surface upon which to adhere and build up a thin layer. Thin coatings of metals and many ceramic substances can be made in this fashion. Sputtering is a similar process, but it uses a high voltage and a radio frequency field to knock molecules from a target material and deposit them onto a material placed at the opposite electrode. This technique is used to deposit refractive materials that cannot be evaporated in a vacuum system.

Another way to form an antireflection coating is to make it out of the top layer of silicon. Silicon can react with oxygen- or nitrogen-containing gases to create silicon dioxide or silicon nitride. Although it is difficult to obtain a sufficiently thick layer of silicon dioxide by this process, silicon nitride antireflection coatings are used on commercial solar cells.

A final refinement common in space-quality cells, and used by at least one manufacturer of terrestrial cells, is the multilayer antireflection coating. Three or more layers of a set of two different transparent substances cancels reflected light almost completely, giving a reflection loss of less than 3%. The loss to bare silicon is 35%.

## ASSEMBLY INTO MODULES

The cell is now completed and ready to use. Most modules are assembled by the individual cell manufacturers; the design of the cell's finger pattern is based on the company's particular module configuration. Chapter 3 describes module characteristics in some detail.

***Figure 2.6.*** Automatic machinery soldering leads onto individual solar cells. These machines can replace tedious hand-soldering and produce a better joint. (Siemens Solar Industries, Camarillo, CA)

***Figure 2.7.*** Assembling solar cells into strings.
(AstroPower, Newark, DE)

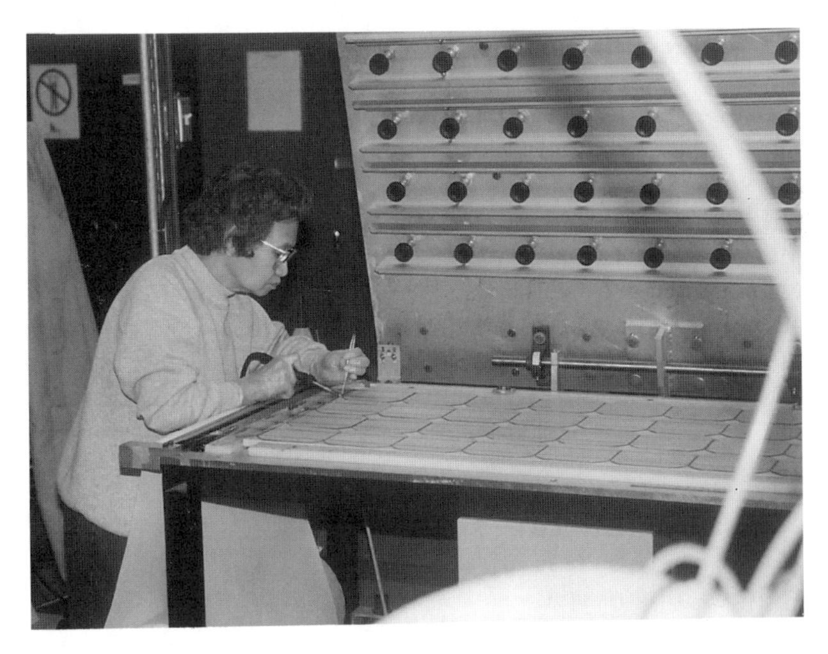

***Figure 2.8.*** Soldering the final connections on a solar cell module. Even in the most automated systems, some handwork is still necessary. (AstroPower, Newark, DE)

***Figure 2.9.*** Sealing the finished module. The best systems are hermetically sealed in a process similar to that used to make safety glass for car windshields. (AstroPower, Newark, DE)

In general, solar cell manufacturers are attempting to institute more and more automation into the module assembly process, which consumes some 40% of the cost of the finished product. Automatic soldering and spot-welding machines are replacing crews that solder cells together by hand. Encapsulant systems with heat-curing plastics have almost completely replaced silicone resins. New simplified wire connectors are replacing expensive junction boxes. The final product is a less expensive but more rugged and reliable device to generate electricity from the sun's energy. The typical PV module has a warranty of twenty years or more.

## NEW PRODUCTION TECHNIQUES FOR SILICON

### Solar-Grade Silicon

The technology for manufacturing silicon solid-state devices has been so well developed, and so much research effort has been invested in the silicon solar cell, that the most logical way to cut costs is to devise new ways to make this cell. A number of people have been doing just that.

Most solar cell manufacturers have automated the assembly lines that manufacture the cells and assemble them into modules. These lines have greatly increased production capabilities and have lowered module costs to the point that distributors are selling complete modules to small users for $5.50 per watt. Manufacturer-direct sales to large projects have had prices as low as $4 per watt.

The ultrapure semiconductor-grade silicon currently used in the manufacture of single-crystal cells is getting more expensive and more scarce as the electronics industry competes with the cell industry for the limited output of the few producers. The question is, do solar cells require starting material of such high quality? The answer is, not really. **Solar-grade silicon** produces 15 to 16% efficiency solar cells, but costs only a fraction of the current $25 per kilogram paid for the **electronic-grade** used in integrated circuits. Wacker Chemitronic has set up a production facility in the United States to produce nothing but solar cell-grade silicon. Also, Jet Propulsion Laboratories, under contract with the Department of Energy, initiated a research program aimed at developing new processes to produce this grade of silicon in a cost- and energy-efficient manner. This program has attained the objective of encouraging some manufacturers to implement new processes.

## Polycrystalline Silicon

One of the most expensive, energy-intensive steps in making a solar cell is pulling the large single crystal that serves as the starting material. If a perfect single crystal were not needed, a great deal of cost could be eliminated. Some researchers who have tested polycrystalline silicon believe that the loss of efficiency caused by the grain boundaries is just too great, but several methods of casting cubes of polycrystalline silicon with large crystal grain sizes have led to polycrystalline solar cells with efficiencies as high as those of the commercial single-crystal cells.

Joe Lindmayer of Solarex started a companion company called Semix, in joint venture with Amoco, to manufacture "semicrystalline" solar cells from large blocks of silicon cast using a proprietary process. The cooling of the molten silicon is controlled so that the impurities in the starting material are swept to one end of the block and then discarded. This process utilizes **metallurgical-grade silicon**, which is much cheaper than electronic-grade. The resulting crystal grains in the blocks are so large that the cells produced behave almost like single-crystal cells. If the grain boundaries are perpendicular to the face of the device, the cell acts like a group of small cells in parallel. In 1983, Semix constructed its "Solar Breeder" plant in Gaithersburg, Maryland, to make this new type of cell directly from high-quality sand. Solar cells on the roof of the Breeder supply some of the power needed to run the operation, which is capable of producing enough cells each year to generate 25 peak megawatts of electric power. Since the takeover of Solarex by Amoco in 1984, the Breeder has been put into operation and is Solarex's main production facility.

Crystal Systems has perfected a similar system in which they can produce a cube of single-crystal silicon 12 inches by 8 inches, weighing 90 pounds (Figure 2.10). Ironically, this brings us back to single-crystal cells, but this crystal-growing process is very energy-efficient and produces high-quality cells from a metallurgical-grade starting material. The process, which involves a helium gas heat exchanger, casts the ingot in a silica crucible which breaks apart upon

***Figure 2.10***. Heat exchanger method (HEM) single-crystal silicon ingot. (Crystal Systems, Salem, MA)

cooling and leaves the cube intact. Other organizations, including Wacker Chemitronic, are casting polycrystalline cubes also, but with smaller grain sizes. Kyocera sells polycrystalline cell modules made by a proprietary casting process.

## Ribbon Growth Systems

The cubes of silicon described above still must be sliced. Although several multiple-wire slicing devices have been developed, the process remains very time-consuming and wasteful. Another approach would be to grow ribbons of single-crystal silicon and cut them to size. The cells would be produced without any further surface treatment. This method saves several steps and is a much more efficient use of expensive pure silicon.

Mobil has been working on ribbon growth systems for some time. Their process involves pulling a vertical ribbon of silicon from a carbon die set in a bath of molten silicon, but there is the problem of inclusions of silicon carbide forming when the hot silicon reacts with the die. The Japanese use boron nitride dies to eliminate this situation.

Mobil's ribbon puller produces a nine-sided cylinder rather than a flat ribbon. The continuous molten circumference allows more careful control over the solidification process and produces high-quality sheets with very few silicon carbide inclusions. The cylinder is sliced into 2-x 4-inch plates using a high-powered laser; solar cells are then produced from these plates through a set of proprietary steps. Mobil sold Mobil Solar to a consortium that includes Germany's Deutsch Aerospace. The new company, ASE Americas, has attained full production. Three members of the old Mobil Solar have started a venture called Evergreen Solar. They have developed a ribbon-growth process that uses a pair of strings to define the edge of the silicon ribbon during the low angle pull of the crystalline ribbon from the silicon melt. Evergreen Solar is in full production also and recently had a public stock offering.

In other ribbon growth methods, the ribbon is pulled horizontally or even downward or no die is used at all. D. N. Jewett of Energy Materials Corporation developed a process that pulls the silicon sheet almost horizontally from a quartz crucible. This system (Figure 2.11) has the advantage of operating at high pulling speeds while producing high-quality, defect-free ribbons. It is not currently in use.

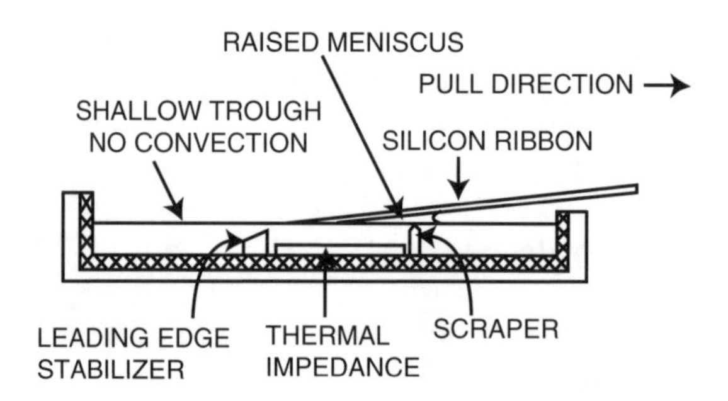


***Figure 2.11.*** Silicon ribbon-pulling process developed by D. N. Jewett. (Energy Materials Corporation, Harvard, MA)

A Westinghouse process exploits the dendritic growth of a pair of crystals at the edges of the ribbon to control the shape; Solavolt International developed a process in which a polycrystalline sheet, produced by chemical vapor deposition from silicon compounds, is melted very quickly by a scanning laser to produce a single-crystal ribbon. This process, shown in Figure 2.12, has an uncertain fate at the moment since Solavolt production facilities have been shut down. In 1982, Honeywell and Coors, in a joint project financed by the Department of Energy, experimented with ribbons of silicon cast onto ceramic substrates, but all work stopped when government funding ran out. In contrast, a group in France continues similar work with a graphite supporting matrix. The silicon layer can be very thin and if the substrates can be produced cheaply enough, the cell could meet DOE's cost objective of $0.70 per peak watt.

AstroPower, a photovoltaic manufacturer in Delaware, has set up a pilot plant making cells from thin films of crystalline silicon coated onto a low-cost proprietary substrate. Although AstroPower has shipped special-project PV modules made by this process, which promises to be inexpensive, they still make commercial solar cells from single-crystal wafers recycled from the electronics industry.

## AMORPHOUS SILICON

One of the most surprising industrial developments has been the rapid growth of the amorphous solar cell. From a laboratory curiosity in the

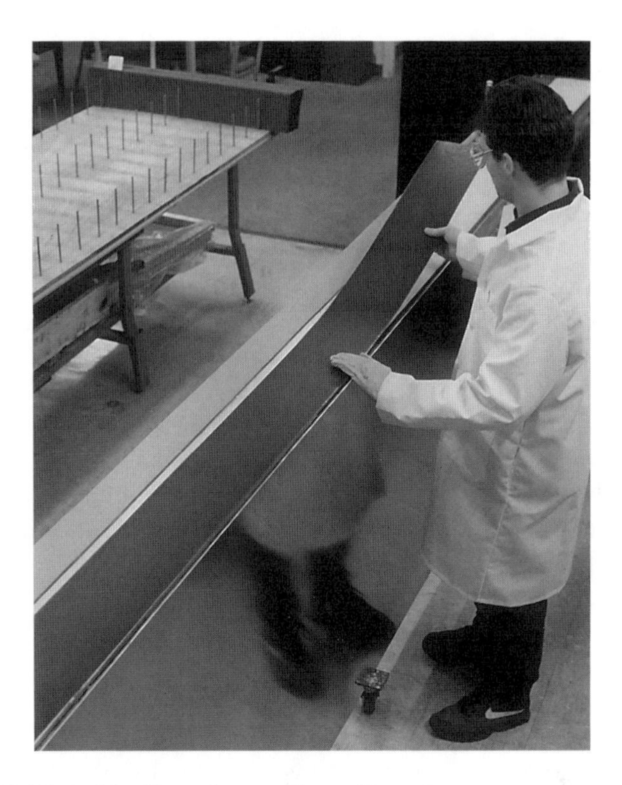

***Figure 2.12.*** A thin film of crystalline silicon is grown on a proprietary substrate to make low-cost solar cells. (AstroPower, Newark, DE)

early 1980s, the cell has become a standard part of pocket calculators and watches. (Industry sales amounted to 100 million calculators in 1994, a substantial segment of the total Japanese cell output.) An extremely sophisticated automated manufacturing process has reduced the price of a set of four or five small cells to less than the cost of the batteries they replace. Even though the cost per watt of small amorphous cells is around $4, the milliwatt power cells in use are only a few cents each.

The two main obstacles to immediate large-scale utilization and production of amorphous power modules are (1) light-induced degradation and (2) poor yield in large-area cell production. To understand this, we need to know more about **amorphous semiconductors**.

An amorphous material is one that appears to be solid but has no long-range crystal structure. Glass is a good example of this. Amorphous selenium has been known as a photovoltaic cell material for over a century and has been used in Xerox photocopiers since the late 1940s. The semiconducting properties and the band structure of this material have been examined, and the photovoltaic cells made from selenium have been used commercially for decades even though they have an efficiency of only about 0.1%.

In 1966 Yoshihero Hamakawa of Osaka University in Japan discovered a way to make **amorphous silicon** solar cells. His work was ignored until the mid-1970s when RCA began investigating amorphous silicon as a potential solar cell material. The material had been known for some time but had been dismissed as a solar electric converter because it normally has a high resistance. Actually, if the amorphous silicon thin film is deposited in such a way as to incorporate a small percentage of hydrogen and other elements trapped in the layer, the high dark resistance is immaterial and the solar cell will have a relatively high short-circuit current. It is believed that the hydrogen atoms form Si–H bonds to tie up some of the dangling bonds caused by the somewhat random arrangement of the silicon atoms in the amorphous layer.

The usual way to deposit the amorphous silicon layer is by **sputtering** silane ($SiH_4$) or some other silicon compound in a vacuum system. The reactive sputtering is achieved by producing a glow discharge in a low-pressure gas mixture using a high-voltage DC or radio-frequency AC power applied between two electrodes. One of the electrodes or a substrate placed between the electrodes will become coated with the amorphous film.

The amorphous silicon cell has developed commercially over the past decade and now constitutes an important branch of the photovoltaics industry.

Amorphous silicon has a band gap somewhat larger than crystalline silicon and absorbs light more strongly in the visible part of the spectrum, which means that the theoretical efficiency of solar cells made of this material is actually higher than that for single-crystal silicon cells. A great deal of work remains to be done to clarify semiconducting properties and efficiency loss mechanisms before this efficiency can be realized. However, the simple fabrication steps and the thin films that can be used in making the solar cells should produce a practical, inexpensive cell even if the production-line efficiency is no more than 8%, although 15% has been achieved already in the laboratory. Questions about the long-term stability of amorphous films under sunlight conditions are beginning to be answered, as the Stabler Wronski effect (named after its discoverers) is becoming understood and thinner multi-layer cells diminish its importance.

Researchers at Osaka University in Japan and elsewhere have been working on a heterojunction solar cell combining amorphous silicon carbide with amorphous silicon. It is made by a sputtering system also, but uses a mixture of silane and methane as the reacting gas. This device has achieved 7.5% efficiency.

The first problem, the deterioration of the cell after exposure to direct sunlight for as little as one day, is poorly understood, even though considerable research effort has been directed toward the phenomenon. The extent of decline has been reduced by changing the formula of the complex mixture of gases used in the sputtering deposition of the amorphous layer, but the effect is still present and, for now, simply accepted. Some manufacturers refer to this as the "burning-in period" and rate performance after degradation. The relatively free movement of small atomic species such as hydrogen and fluorine in the loosely linked silicon atom matrix could be responsible for this Stabler Wronski effect; a good deal of research is aimed at replacing the fluorine and "locking up" the matrix with larger atoms, such as other halogens and boron. Boron, of course, also acts as a p-type dopant, giving the amorphous film the proper electronic characteristics for a solar cell. The research currently underway worldwide will certainly pay off in the near future and we will soon see commercially available amorphous solar cells of 8 to 10% efficiency. In fact in 1994, the National Energy Research Laboratory (NERL) announced a record performance of 10.8% after "stabilization" of an amorphous silicon multilayer solar cell manufactured by United Solar Systems (a division of Energy Conversion Devices) of Troy, Michigan. The company is now producing inexpensive "solar roof shingles" using this technology.

The second problem, poor yield in large-scale manufacture, could simply be one of quality control, similar to the problems that plague all semiconductor manufacturers when they enter a new product area. It is possible, though, that the point defects that could short out an entire cell will prove impossible to eliminate. In this case, the best strategy might be to institute an "integrated circuit" approach—large

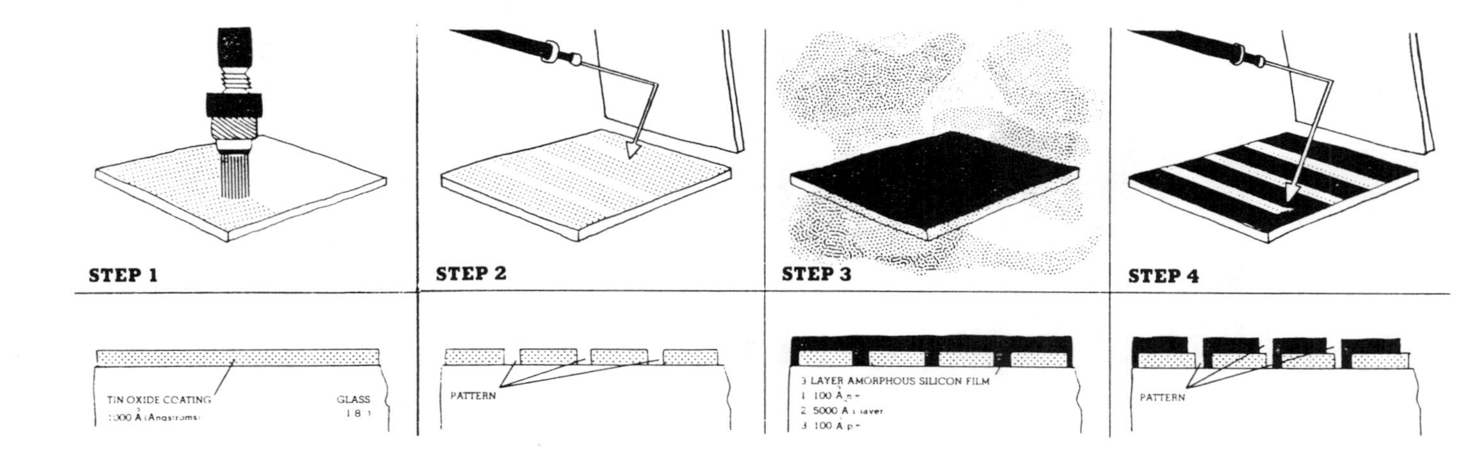


**STEP 1** The glass substrate is coated with a transparent conductive coating. Tin oxide is deposited by a chemical spray deposition process from organometallic tin compounds.

**STEP 2** The transparent conductor is patterned into regions of conductive and nonconductive areas by removing the tin oxide coating with a scanned high power laser.

**STEP 3** A deposition chamber lays a film of amorphous silicon on top of the patterned tin oxide. Three different layers of amorphous silicon are deposited: one layer with p-type conductivity, a middle intrinsic layer, and a top layer with n-type conductivity. This configuration converts sunlight into electricity.

**STEP 4** Strips of the amorphous silicon films are removed, using laser techniques similar to Step 2.

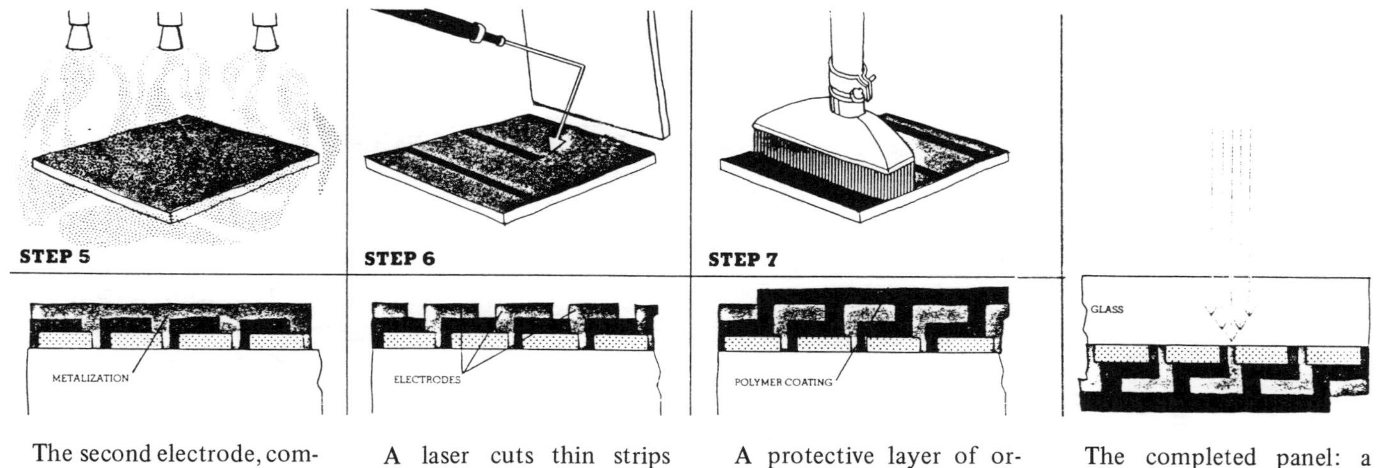


The second electrode, composed of a double metal layer (one is aluminum), is deposited in an evaporation chamber. The evaporated metals are deposited in a thin film onto the scribed amorphous silicon.

A laser cuts thin strips from the metal electrode.

A protective layer of organic polymer is deposited on the back of the panel, using a combined dip/spray process.

The completed panel: a number of cells connected in series and parallel. The two output conductors are strips of copper tape on opposing edges. The supporting frame of the array contains the inter-panel connectors.

***Figure 2.13.*** The step-by-step process of amorphous silicon photovoltaic panel manufacture. This process is capable of handling glass up to one foot by three feet. By varying the laser scribing pattern of the various layers, different combinations of voltage and current can be obtained. (Advanced Photovoltaic Systems, Princeton, NJ)

amorphous photovoltaic modules composed of thousands of tiny cells all made at one time on a single substrate, connected only after testing has revealed the defects.

It is becoming difficult to keep track of the corporations that have announced new or expanded efforts in the amorphous solar cell area. The Japanese are still slightly in the lead in production output and, either directly or through joint ventures with American companies, such as United Solar Systems, are responsible for most current commercial sales (although U.S. researchers are leading in new developments). Japan has built demonstration homes with amorphous power panel roofs and several companies have shown prototype power panels in the 30- to 40-watt size. None of these companies accept orders for immediate delivery; schedules are vague and have a tendency to slip. Actual full-scale production will have to wait until these problems are solved.

In the United States, United Solar Systems and Advanced Photovoltaic Systems, Inc. (formerly Chronar), of Princeton, New Jersey, depend on amorphous cells for their entire product line. United Solar Systems, in conjunction with Sharp, maintains a full-scale automated production line in Japan to fabricate a wide ribbon of cells on a continuous basis; two more lines will be operable soon in the U.S. and the technology has been exported on a turn-key basis to Egypt. While United Solar Systems has demonstrated large-sized, flexible amorphous modules, their production modules are small units for calculators and consumer devices that are not exposed to long hours in the direct sun.

Chronar, on the other hand, concentrated on large-wattage power modules from the start and consequently was slow to sell solar products on the open market. Chronar's profits had been produced by selling technology—arranging joint ventures in Egypt and elsewhere to set up amorphous assembly plants. Their own plant, built in conjunction with AFG Industries, Inc., used automated technology to build amorphous modules onto tin oxide–coated glass (see Figure 2.13). These small cells, produced for use in calculators, run at about 3% efficiency. Chronar also introduced an innovative solar patio light which used a 2-watt amorphous silicon panel to charge a battery that operates a small (0.8 watt) light. Although these inexpensive outdoor lights sold very well, losses soon forced a reorganization of the corporation a few years later, with Advanced Photovoltaic Systems assuming the photovoltaic business.

Most major solar cell manufacturers have announced amorphous silicon research and development efforts. Some companies, like Exxon and Photowatt SA, have closed their crystalline silicon operations, pinning all their hopes on a future amorphous solar cell line. Others, like Siemens Solar (formerly Arco Solar) and Solarex (now part of BP Solar) are phasing in amorphous products as they are ready, but are not decreasing conventional solar module production. Solarex, for example, bought RCA's amorphous silicon research and development section, as well as the U.S. patents for amorphous silicon cells. In fact, it was the pioneering work at RCA, the first American company to work on amorphous silicon cells, that led to this recently developed industry. Even though the basic cell was discovered in a Japanese university laboratory, Japanese companies paid no attention to the work until RCA announced its amorphous program at the 1976 IEEE Photovoltaic Specialists Conference.

The boom in amorphous silicon solar cells will continue. It is only a matter of time before the technical problems will be solved and large power modules become generally available. The price per watt for these devices, however, will not drop precipitously. The cost of manufacture and the high reject rate will make it difficult, at first, to profit, especially in competition with the present price per watt of crystalline silicon cells. The price will start to decline only after large-volume deliveries have given production people the time and experience to take advantage of the inherent economics of amorphous devices. As this happens, new markets will open up, and amorphous photovoltaics will begin to produce a significant amount of our electric power.

## RECOMMENDED READINGS

Chalmers, Bruce. "The Photovoltaic Generation of Electricity." *Scientific American* 235:34–43 (October 1976).

Hubbard, H. M. "Photovoltaics Today and Tomorrow." *Science* 244: 297–304 (April 21, 1989).

Perez-Albuerne, E. A., and Y.-S. Tyan. "Photovoltaic Materials." *Science* 208:902–7 (May 23, 1980).

# Chapter 3
# Solar Cells and Modules

## SERIES AND PARALLEL STRINGS

The maximum useful voltage generated by a solar cell is on the order of 0.5 volt. This is not a very high potential and not many devices can operate on such a small voltage. However, solar cells can be connected in **series** (the back contact of one cell connected to the front contact of the next) to obtain a higher voltage. Figure 3.1 (top) shows such a series string, which is called a **solar module**. (Although the terms **solar array** and solar module are sometimes used interchangeably, an array is a set of one or more modules.) A typical solar module will have 33 cells in series, producing an open-circuit voltage in bright sunlight of around 20 volts, or 16 volts when producing its maximum power. Such a module is designed to charge a 12-volt storage battery. It is important to remember that the total current output of a series string of cells is the same as that of a single cell.

To increase the current output, solar cells are wired in **parallel** (the back contact of one cell connected to the back contact of the next, with the front contacts similarly connected). In a parallel sys-

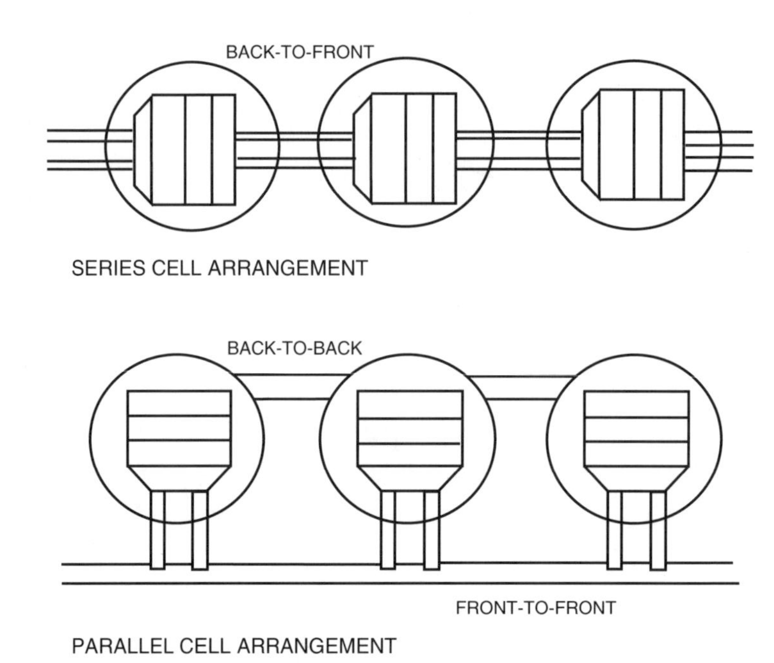


***Figure 3.1.*** *Top:* Series string of solar cells increases voltage. *Bottom:* Parallel string of solar cells increases current.

tem, such as shown in Figure 3.1 (bottom), the total current is the sum of the individual current outputs of the cells, but the total voltage is the same as the voltage of a single cell. Usually, individual cells are wired in series, forming a module, to obtain the desired output voltage; modules are wired in parallel in an array to obtain the desired current. Chapter 4 details this procedure.

Solar cell manufacturers produce a variety of modules that can be combined into an array to meet just about any voltage and current need (see Figure 3.2).

***Figure 3.2.*** Finished AstroPower solar cells exiting the furnace where the top finger contacts were fused onto the single crystal silicon. (AstroPower, Newark, DE)

## MODULE CONSTRUCTION TECHNIQUES

Since silicon solar cells and arrays were first developed for the space program, the construction techniques emphasized reliability. In most cells the front contact finger patterns are designed with some redundancy so that a break in the pattern or a bad solder joint will not shut down the entire cell. The AstroPower cells, shown in Figure 3.3, illustrate this design technique. The cells are fastened into series strings by sweat-soldering thin, tin-plated copper ribbons over the three main crossbars. Then the ribbons are sweat-soldered onto the back of the next cell. Even if a

***Figure 3.3.*** These single crystal silicon solar cells have efficiencies as high as 18%. (AstroPower, Newark, DE)

cell cracks, the two pieces are still fastened in parallel and electrical performance is virtually unchanged.

The cell pictured here illustrates the older technique of nickel fingers tinned by an industrial wave-soldering machine. Other cells, primarily used for space applications, are made with silver-plated and palladium-silver fingers. Silkscreening techniques are now almost universally used for large production terrestrial cells. Corrosion and mechanical failure of the finger contacts and interconnects are the most common causes of cell failure, so noncorrosive metals must be used. However, the metals must have the proper electronic behavior in contact with silicon. Most cells are soldered together to make modules, but some manufacturers now spot-weld the interconnections. Many small solar modules come equipped with a built-in **blocking diode**. This diode protects a storage battery against drainage by preventing the current from running backward through the solar cell array. Blocking diodes will be discussed further in the next chapter.

## ENCAPSULATION

Cells must be sealed to protect them from the environment and to support them in the module. The two most common encapsulants are transparent silicone rubber and an ethylene vinyl acetate (EVA) plastic similar to the plastic layer used in automobile safety glass. Usually a top cover of plastic or tempered glass is added to offer better protection against the elements. Glass covers are more scratch-resistant and remain transparent longer, but they do not flex as much as plastic covers. This flexibility makes plastic modules more suitable for use on sailboats and camping vehicles. Of the two plastics most commonly used, polycarbonate is the more damage-resistant and can withstand higher temperatures, but acrylic is more flexible and less expensive.

Bringing the electrical wires out of the panel requires careful design. The electrical connectors are a pair of contacts that protrude from the back of the panel and are sometimes encased in a small junction box. Some modules are designed so that the wires project from the side, which allows the panel to rest flat on a support surface (for example, a roof or sailboat cabin top). These connectors must be very well sealed, so that water cannot seep in along the wires and corrode the interior contacts. They also must be very secure so that the wires cannot be pulled out or put strain on the cells.

## MODULE FAILURE MECHANISMS

### Joints Between Cells

The silicon solar cell is a remarkably rugged device. The cell itself can withstand a great deal of abuse and extreme temperatures, both high and low. It is also impervious to most corrosive chemicals. When a solar module fails, the problem usually lies with something other than the cells.

The first solar module I constructed failed prematurely after four years of operation on the deck of a small sailboat. The main problem was the corrosion caused by saltwater that had leaked into the case. Some of the joints between cells had failed mechanically, as well. The cells had been fastened in series by a technique called **shin-**

**gling**. The front edge of one cell is overlapped by the back edge of the next, similar to roof shingles, and the edges soldered directly together. No copper jumpers are used to interconnect the cells except from one row to the next. Since the entire row is one rigidly fastened unit, any flexing or thermal expansion puts a great strain on the adjoining edges, pulling the top contact loose from the underlying cell.

Producing a good mechanical bond between the silicon cell material and the nickel or silver top finger contact is difficult, and it is surprisingly easy to pull the fingers right off the cell surface by jerking on the jumper strip soldered to the fingers. However, because the newer silk-screened fingers adhere very well, these cells will break before the tabs can be pulled off. The back contact covers the entire back of the cell and, while it is not actually any stronger, the greater surface area helps make the cell more durable.

Too high a temperature during the process of soldering contacts to the cells may also loosen the fingers from the silicon; too low a temperature will cause a "cold" solder joint that will quickly fail. The exact composition of the solder and the flux, and the cleanliness of the operation, greatly influence the reliability of the joints.

### Encapsulant Problems

If the thermal expansion of a rigid encapsulant differs sufficiently from that of the cells, the encapsulant can pull loose or the cells can crack. This has been a problem with cells embedded in polyester resins.

Direct sunlight creates a very harsh environment and can cause chemical reactions in encapsulants and cover materials, turning them cloudy or yellow. While this may have no effect on the cell itself, it will cut down on the amount of light that reaches the cells and this will lower system efficiency. These reactions are accelerated by the high temperatures modules can attain under full sun in the summer. The cells can withstand much higher temperatures than any other part of the module, although cell efficiency drops when the temperature rises. It is best to keep the cell temperature below 60°C.

Recently, some problems have come to light with older PV modules that used EVA encapsulants. After long exposure to high temperatures, the material turns brown and begins to decompose, forming acetic acid that corrodes the interconnect ribbons between cells. This problem occurs most often when reflectors have been added to the system to increase light intensity and power output, but without adequate provision for cooling. Even then, it takes years for this condition to appear.

## Shunt Diodes

A cell can be permanently damaged if a large reverse voltage is applied to the electrodes. A solar cell is an electrical rectifier which means that it passes current in one direction only. In the dark, the silicon solar cell acts just like any other silicon diode rectifier. However, in the manufacture of solar cells, no real attempt is made to build in an ability to withstand a large reverse bias since the cell is not usually used in this way. There is one circumstance under which this reverse voltage condition can occur: when one cell is shaded while the rest of the cells in a series string are in sunlight. When this happens, the current though the string immediately stops and the sum of all the open-circuit voltages of all the other cells shows up across the shaded cell. If the cell cannot withstand this voltage, it will break down electrically and begin to conduct. The resistance heating effect of the current can make a cell hot enough to melt the solder connections and destroy the fingers.

Sooner or later, such shading will occur: for example, a leaf falls on one cell or a tree casts a shadow on a module. Since it is impossible to prevent such occurrences, it is necessary to take precautions so that shading will not destroy a cell. With 33-cell modules, reverse bias breakdown is rarely a problem as virtually all cells will take 30 volts or more in reverse bias. But if higher voltage strings are needed, shunt diodes should be installed. Chapter 4 describes how to wire shunt diodes, as well as other practical details of connecting solar modules. Some manufacturers have constructed experimental cells with the shunt diode built into the cell structure and some amorphous silicon modules have shunt diodes built into each long row. With these cells and modules, no external shunt diode is necessary.

In spite of the potential troubles described above, solar cell modules, if properly designed and installed, can last a remarkably long time. The first solar modules made by Bell Laboratories and originally installed in Americus, Georgia, in 1957, are still in operation. A minimum of 25 years of reliable operation can be expected from a solar electric installation.

## CONCENTRATOR SYSTEMS

One way to use expensive solar cells more efficiently is to concentrate more light onto them. A detailed description of the numerous schemes that have been devised to accomplish this is beyond the scope of this work. Instead, we will discuss the general principles and give examples of each type.

The first parameter that can be used to characterize a concentrator system is the **concentration ratio**. This is simply the ratio between the area of the clear aperture, or opening through which the sunlight enters, and the area of the illuminated cell. Figure 3.4 shows a simple lens-type concentrator similar to a magnifying glass used to start fires. The concentration ratio is the area of the lens divided by the area of the cell. If the lens is 200 mm in diameter, its area is 314 $cm^2$; if the cell is 50 mm in diameter, its area is 20 $cm^2$ (for a circle, the area $A = \pi r^2$). The concentration ratio is 314/20 = 16 to 1 (sometimes expressed as a concentration ratio of 16 suns). This means that the cell receives 16 times the light (minus the reflection and absorption losses in the lens, which can be very small for a coated lens) and should put out 16 times the power of a similar cell used without the lens. Because the lens does not have to be of fine optical quality (it could be molded plastic), it would seem that this system would be much cheaper than the 16 cells it replaces. The reality is not quite so simple.

First, with all the extra sunlight pouring onto it, the cell will heat up. With a concentration ratio of three suns or more, some way must be found to cool the cell. Simple cooling fins mounted to the back of the cell may suffice, but in hot climates or with higher concentration ratios, water or forced-air cooling becomes necessary. The usual EVA encapsulant is unsuitable for concentrator systems because it degrades at the higher temperatures and light intensities, so the more expensive silicone encapsulant is generally used. Silicone RTV has an unusual combination of properties in that it is an excellent electrical insulator while it conducts heat reasonably well and it can withstand high temperatures without yellowing; the excess heat removed

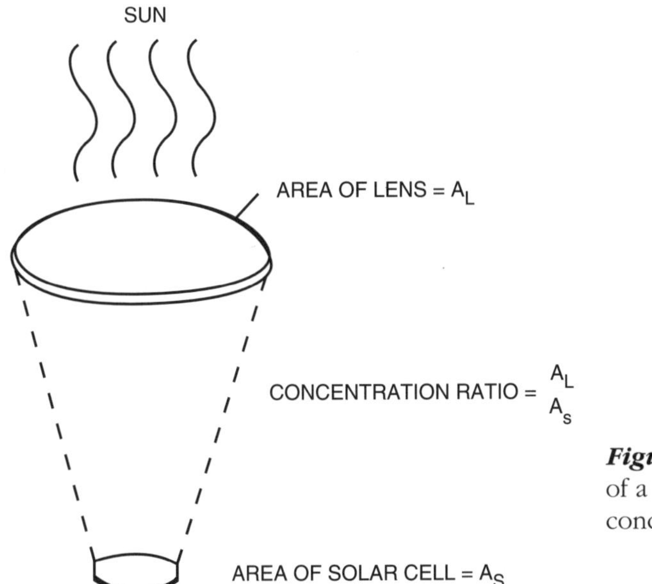


***Figure 3.4.*** Diagram of a simple lens-type concentrator.

from the cells can be saved and used. A system that produces usable heat as well as electricity is a **hybrid system**. Hybrid systems can be very cost-effective when neither the heat nor the electricity generated separately would justify a solar system. Hybrid systems are discussed later in this chapter and again in Appendix A, which contains step-by-step instructions for hybrid module construction.

Second, the currents produced by the photovoltaic cells in concentrator systems can become quite large. (Currents in excess of 100 amps have been produced by a single cell.) In order to handle these massive currents, special cells have to be constructed. Usually these cells are thinner than the normal cell and have a much smaller cell resistance. The finger pattern and top contact for the cell must be carefully designed to carry these currents without blocking too much of the cell's surface area. Figure 3.5 shows a concentrator cell capable of handling a concentration ratio of up to 200 suns. Naturally, such cells are more expensive than the standard solar cell, but each one produces so much power that the extra cost is justified. Ordinary solar cells can handle a concentration ratio of 3 or 4 suns, provided they are cooled adequately.

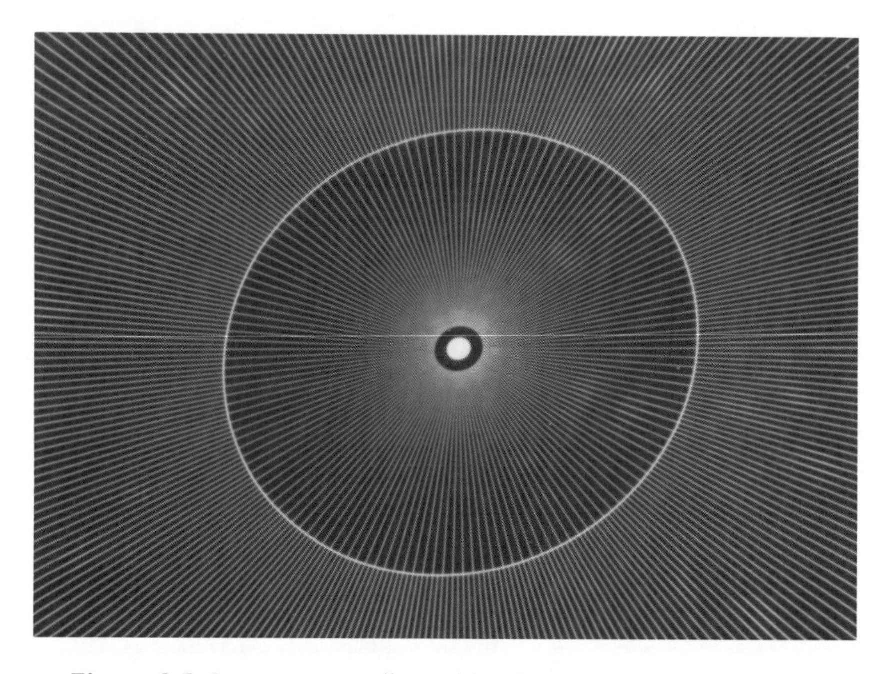

***Figure 3.5.*** Concentrator cell capable of handling ratios of 200 suns. (Applied Solar Energy Corporation, City of Industry, CA)

Third, for a concentrator system to work best, it must point directly at the sun at all times. Therefore, a tracking system must be devised to pivot and follow the sun as it travels daily from east to west. The pivot must also raise and lower the angle to allow for the seasonal change in the sun's path in the sky—high in the summer and low in the winter. The simple concentrator shown in Figure 3.4 now has a complex tracking mount. The mount could be moved by hand, which would require constant adjustment, or it could have a clockwork mechanism that tracks the sun automatically. The most sophisticated systems have photocell detectors and feedback mechanisms to constantly correct the system and keep it directed correctly.

## Two-Axis Tracking

A mount that is capable of pivoting both daily and seasonally to follow the sun's path is called a **two-axis tracking system**. Many two-axis tracking systems have been built experimentally and several are available commercially. Some units use a small concentration ratio of 3 to 1, so a simple cone reflector is sufficient and large tracking errors can be tolerated. If the unit also has forced-air cooling of the cells, the hot air produced can be put to some useful purpose. Such modules were commercially available back in the 1980s.

Figure 3.6 shows another two-axis tracking system that uses **Fresnel lenses** as the concentrator elements. Fresnel lenses work as well for light concentration as the thick lenses they replace, but they use only a fraction of the material and can be molded out of plastic. It is also possible to have a greater optical correction in the outer edge of a Fresnel lens so that the distance from the lens to the cell can be shorter for the same concentration ratio. This particular Fresnel lens concentrator system, produced by Martin Marietta Corporation, was installed in a village in Saudi Arabia and has a 350-kW capacity, enough to supply the power needed by two villages. (It replaced a set of diesel generators.) With a concentration ratio of 33 suns, the system has a particularly sophisticated tracking controller using coarse and fine sun sensors and a small computer. The system is passively cooled and no provisions were made to collect the waste heat.

Another commercially available two-axis concentrator is being manufactured by Midway Labs in Chicago. This system, developed by Roland Winston of the University of Chicago, uses two-stage concentrator optics to reach a concentration ratio of 335 to 1. Complete with tracking mount, it is cost-competitive with a fixed array of flat PV modules of equivalent electrical output.

### One-Axis Tracking

The purpose of a solar concentrator is not to give a clear image of the sun, but rather to collect the sunlight that falls over a large area and project it onto the smaller area of the collector or array. The quality of the image does not matter as long as most of the sunlight falls onto the cells. Figure 3.7 shows a trough-type parabolic collector that does this in one dimension. This particular embodiment of the trough collector is called a **Winston Concentrator** as it was also developed by Roland Winston. Its great advantage for smaller concentration ratios is that it does not have to track the sun daily for the seasonal variation in the sun's position. Table 3.1 gives the acceptance angle and frequency of adjustment for different concentration

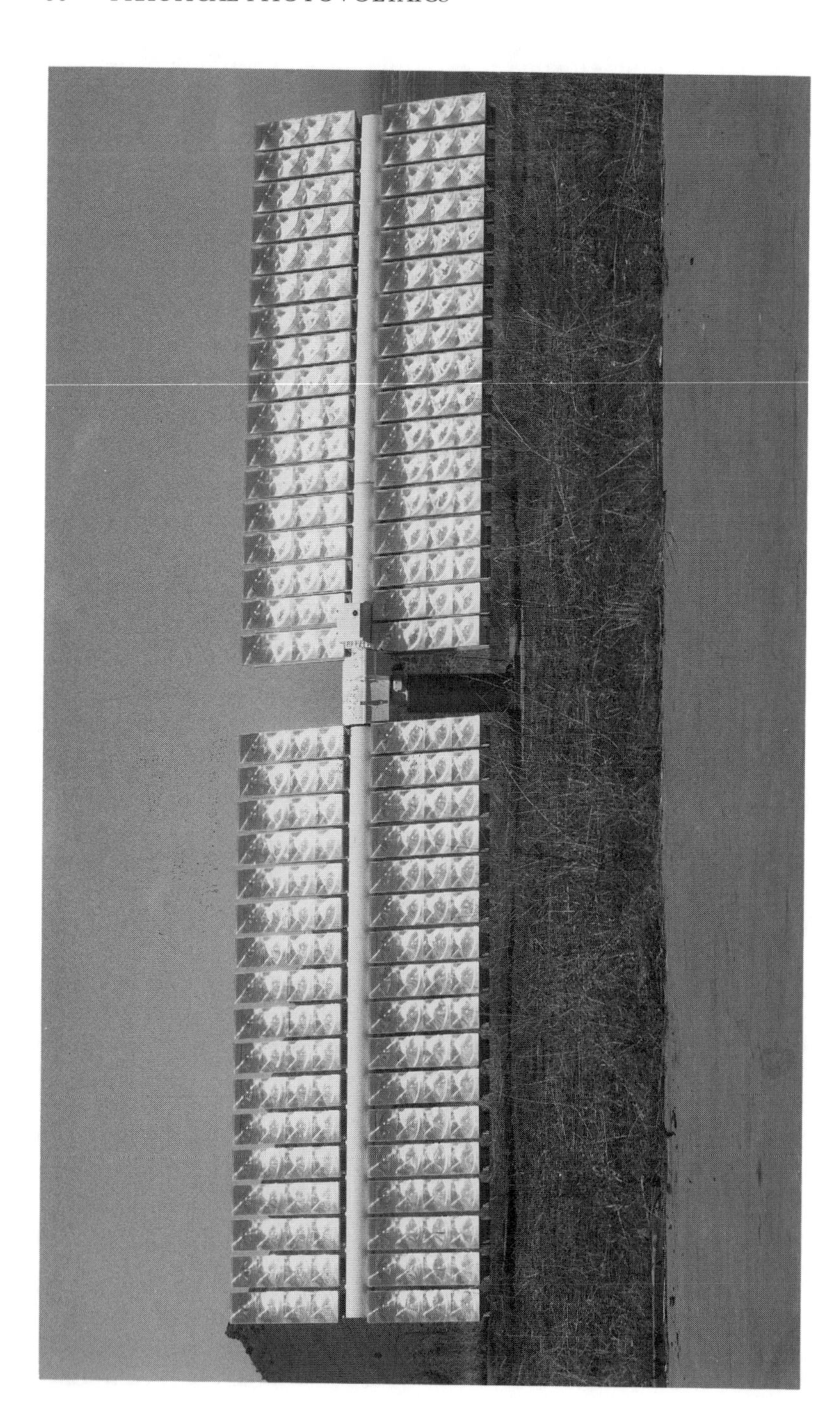

***Figure 3.6***. Fresnel lens two-axis tracking system array. (Martin Marietta Aerospace, Denver, CO)

ratios. The **acceptance angle** is the total range of sun positions over which the sunlight will still be collected by the mirror system and focused onto the solar cells. This table assumes that the axis of the trough collector is horizontal and pointing east and west.

This simple system does not effectively collect sunlight early in the morning or just before sundown when the sun is shining more or less lengthwise down the collector. Another version of the linear concentrator is oriented north and south and tilted south at an angle equal to the latitude of the installation. The system tracks the sun during the course of a day but needs no adjustment for seasons and so the mount and tracking mechanism can be quite simple. Also, connecting the cooling fluid pipes or air ducts to the system is greatly simplified if a hybrid or **total energy system** is desired.

***Figure 3.7.*** Trough-type parabolic collector, similar to that developed by Winston, mounted on the roof of the Raymond Frady residence near Mifflin, Indiana. This type of hybrid concentrator photovoltaic system is particularly suitable for remote residential applications.

***Table 3.1***
**Acceptance Angle and Adjustments Needed per Year for Different Concentration Ratios for a Winston Concentrator**

| Concentration Ratio | Acceptance Angle (degrees) | Number of Adjustments per Year | Shortest Period between Adjustments (days) |
|---|---|---|---|
| 2 | 30.0 | 0 | — |
| 3 | 20.0 | 2 | 180 |
| 4 | 14.5 | 4 | 35 |
| 6 | 9.6 | 10 | 26 |
| 10 | 5.8 | 82 | 1 |

Whatever configuration is used, flat plate or concentrator, hybrid or electricity-collecting only, the inherent simplicity of operation makes a PV system easy to install and maintain. The only connections to the solar module are a pair of wires, giving the user a great deal of design flexibility. The next chapter explains exactly how solar cells are used.

## RECOMMENDED READINGS

Backus, Charles E., Ed. *Solar Cells* (New York: Institute of Electrical and Electronics Engineers, 1976).

Commoner, Barry. "The Solar Transition." *The New Yorker* 55:53–54, Part I (April 23, 1979); 55:46–48, Part II (April 30, 1979)

Fan, John. "Solar Cells: Plugging into the Sun." *MIT Technology Review* 80:14–15 (August 1980).

Institute of Electrical and Electronics Engineers. *Spectrum* 17:(February 1980). A special issue devoted to photovoltaics.

Merrigan, Joseph A. *Sunlight into Electricity* (Cambridge, MA: The MIT Press, 1975).

Perez, Richard. "The Fire Within." *Home Power* 40:28–31 (April/May 1994).

Pulfrey, David, Ed. *Photovoltaic Power Generation* (New York: Van Nostrand Reinhold Co., 1978).

# Chapter 4
# Using Photovoltaics

Solar cells are not difficult to use. In this chapter, we will outline how to determine what type of photovoltaic system is needed and then examine the practical details of system installation and operation.

Before deciding on photovoltaics, you need to compare the cost of a PV system to the costs of the alternatives. Solar cells are expensive. Remember, however, that after the initial expenditure for a system is made, utility bills are low or nonexistent and maintenance costs are minimal. (And remember that not all "costs" can be calculated in dollars and cents.) In addition to the traditional PV applications—remote locations where utility power is unavailable, campers and boats, and for temporary power needs from disaster situations to laptop computer battery packs—solar cells are now routinely providing site-specific energy for urban and suburban homes, office buildings, and a multitude of mainstream, grid-connected purposes. Understandably, photovoltaic systems have become very important sources of energy in the developing world. For an increasing number of power needs, photovoltaics is the cheapest and best way to generate electricity.

## THE CURRENT COST OF SOLAR ELECTRICITY

To give you an idea of what solar electricity costs now, here are sample calculations based on either: (a) the best purchase price by the federal government to date ($3.75 per peak watt) or (b) the best purchase price by a private individual for a single finished module (about $5 per watt).

For a person, in Indianapolis using the weather data from the Motorola analysis, one watt of solar cell capacity would produce about 1.2 kWh of electricity a year. (In Arizona, for comparison, one could expect 50 kWh per year per watt.) Over the 30-year life expectancy of a module (they will last much longer, but this is the standard expected lifetime), the system will produce 1.2 kWh x 25 years = 30 kWh. At $3.75 per peak watt, the electricity costs $0.125/kWh; at $5 per peak watt, the cost is $0.17 per kWh. This does not include the cost of other system components, such as support structures, wiring, or storage batteries, which could easily double the final price. However, as solar cell prices continue to drop and utility prices continue to climb—along with concerns for the environment—it will become increasingly evident that solar electric systems provide a safe, sane, and ultimately economically wise solution.

* * *

The photographs on the following pages show a few of the many ways solar modules are being used. For an in-depth review of solar cell use throughout the world, see John Perlin's *From Space to Earth: The Story of Solar Electricity*, published by **aatec publications**.

***Figure 4.1.*** A stand-alone PV system that operates a pump to supply water to cattle. (New Mexico Solar Energy Institute, Las Cruces, NM)

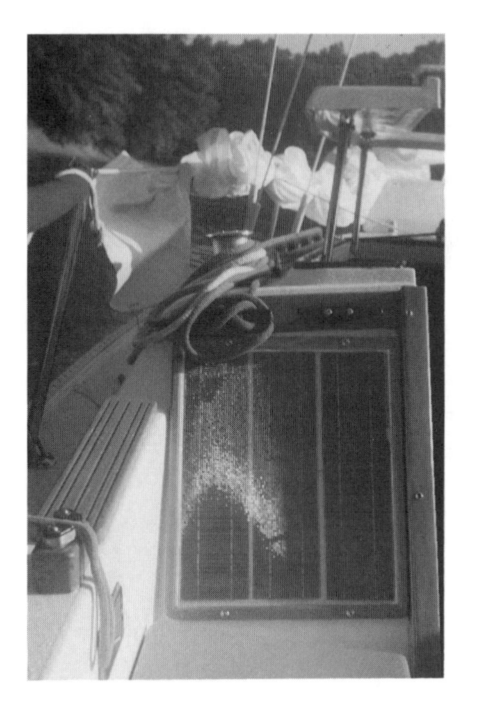

*Figure 4.2*. Teak-framed solar panels mounted on a Catalina 25 sailboat. (Allen Quillian photo)

***Figure 4.3.*** *Solar Challenger*, designed and built by a team headed by Dr. Paul Macready, is capable of 100-mile flights powered solely by solar energy. Over 15,000 solar cells are wired into its wings and tail. (E. I. du Pont de Nemours and Company, Inc., Wilmington, DE)

***Figure 4.4.*** A radio-operated navigation station on the west coast of Greenland. Its ni-cad batteries have sufficient capacity to ensure wintertime operation—an important consideration north of the Arctic Circle which is in total darkness for several winter months. (Solarex Corporation, Frederick, MD; KKF Energi, Denmark)

***Figure 4.5.*** Photovoltaic-powered national park outhouse. (Motorola Semiconductor Group, Phoenix, AZ)

***Figure 4.6.*** WBNO, a photovoltaic-powered radio station, near Bryan, Ohio. (Jet Propulsion Laboratory, California Institute of Technology)

***Figure 4.7.*** The *Destiny 2000*, an electric car built by Solar Electric Engineering, comes standard with photovoltaic panels. (Solar Electric Engineering, Inc., Sebastopol, CA)

***Figure 4.8.*** The first solar-powered electric vehicle charging station in the United States. The College of Engineering at the University of South Florida, in cooperation with the Florida Power Corporation, designed this utility-interconnected solar array that can charge batteries for up to twelve vehicles at a time. The excess electricity is sent to the local utility grid. (Siemens Solar Industries, Camarillo, CA; M. Hutton & Company, Atlanta, GA)

***Figure 4.9.*** The U.S. Coast Guard has installed tens of thousands of solar modules on navigation aides, such as this buoy in Los Angeles Harbor. (Siemens Solar, Camarillo, CA)

***Figure 4.10.*** A mobile solar vaccine refrigeration unit in the Sahara Desert. (Siemens Solar Industries, Camarillo, CA)

## SIZING THE ARRAY

### Calculating Load

The first step in determining which solar array to use is to decide what you want it to do. This means that you must calculate the load that the PV system is to power. To do this, you must first ascertain the voltage that your load needs. This is easy for portable or marine systems because they operate from a **DC** supply with storage batteries and the main objective is to keep the batteries charged. A marine system will most likely be 12 volts, although some larger boats have 24- and even 120-volt **AC** power. For our example, assume a 12-volt system. As mentioned earlier, several manufacturers make 32- to 36-cell modules designed to charge a 12-volt battery system. These systems come in a wide range of current outputs and physical sizes.

The next step is to calculate how much power the system will need. This can be expressed in ampere-hours (AH): one ampere hour equals the current of one ampere running for one hour. You must find out or estimate how much current is drawn by each device you want to use and then estimate how long, on average, each device will run. For example, if a lamp draws 2 amps and you expect to use it for 3 hours each night, it will use total of 6 AH (2 amps x 3 hours) of electricity per day. If the array is to power just one or two devices, this is a fairly simple procedure; for a summer cottage or sailboat, however, such calculations may prove complicated and it may be impossible to predict exact usage.

There is a shortcut you can use if you are installing the solar array into an existing 12-volt system that presently runs from storage batteries. Suppose you notice that the battery system runs for about three days without any charging before the voltage gets unacceptably low. This means that you have probably discharged the batteries to 75% of their rated capacity. Find out what the ampere-hour capacity of the batteries is and take 75% of that capacity as the ampere-hours used in the time period. If the batteries in this example are rated at 100 AH, the calculation would be 0.75 x 100 or 75 AH of power over three days. This means you will need 25 AH per day.

At some point, you must decide what type of conventional backup to use and what percentage of the total requirements the PV system will furnish. Operating a large electrical system totally by the sun is a costly matter. The PV system would have to be oversized to compensate for long cloudy spells or for the one time that the expected load is slightly larger than the expected sunlight. The storage system would have to be large enough to take care of the last few percent of demand. In a small system, in which a reasonably sized battery system can operate the load for weeks, the vagaries of weather average out and the sun can be relied on to furnish all the power. Lighted navigation aids and small radio repeaters can operate this way. *The New Solar Electric Home* discusses this and other aspects of installation and use in much more detail.

Once the average number of ampere hours per day has been determined, the next step is to select an array that is large enough to furnish the average amount of power needed. The average amount of sunlight available in different parts of the world varies widely and so must be taken into consideration along with the average amount of light available on different days of the year.

Researchers at the Motorola Semiconductor Group performed a computer analysis of these factors and devised a simplified procedure that will give a good estimate of the system size needed for a particular application. Their step-by-step method follows.

## Sizing Methodology

Matching an electrical load to a photovoltaic array is an involved process that is greatly simplified by using a computer program. The following procedure is based on computer analysis.

### *Step 1*

Determine the daily load ampere-hour requirement by multiplying the load current in amperes by the hours used per day. For a system that operates in more than one mode during the day, for example, transmit or standby, calculate the ampere hours per day required for each mode and add the resultant figures.

*Example:*

| | | |
|---|---|---|
| Transmit 10 amps for 2.5 hr/day | = | 25 AH/day |
| Receive 1.3 amp for 9.5 hr/day | = | 12 AH/day |
| Standby 0.5 amp for 12 hr/day | = | 6 AH/day |
| Total | | 43 AH/day |

Note: Array size and array cost are directly proportional to load!

***Step 2***

Select desired safety factor:

10% for an attended site
15% for an unattended, accessible site
20% for an unattended, inaccessible site
30% for an unattended, inaccessible site of a critical nature

The safety factor accounts for variables and losses in the system. Weather data is the greatest variable since solar irradiation varies from year to year and from site to site. Locating the PV system even 30 miles from the weather reporting station could significantly change the solar irradiation received. In addition, the accuracy of the recording equipment is typically no better than 5 to 10%. Dust will accumulate on the modules between rain showers and reduce output up to 8% (after each rain, full output is restored). Other losses include self-discharge of batteries and bias current of voltage regulators.

***Step 3***

Multiply the daily ampere-hour load as defined in Step 1 by one plus the safety factor selected in Step 2.

*Example* (using 15% safety factor):

43 AH x 1.15 = 49.45 AH (use 50 AH as the system load)

***Step 4***

Find the installation site on a location code map. Use Figure 4.11 for the continental United States and Figure 4.12 for all other worldwide locations. Note the location code for the site.

***Step 5***

From the system output (AH/day) defined in Step 3 and the location code defined in Step 4, use the appropriate multiplier to determine array size in watts:

Area A AH/day x 3.08 = array watts
Area B AH/day x 3.77 = array watts
Area C AH/day x 5.20 = array watts
Area D AH/day x 6.90 = array watts
Area E AH/day x 9.77 = array watts

*Example* (use 50 AH/day and Laramie, Wyoming—Area B):

array watts = 50 x 3.77 = 188 watts (peak)

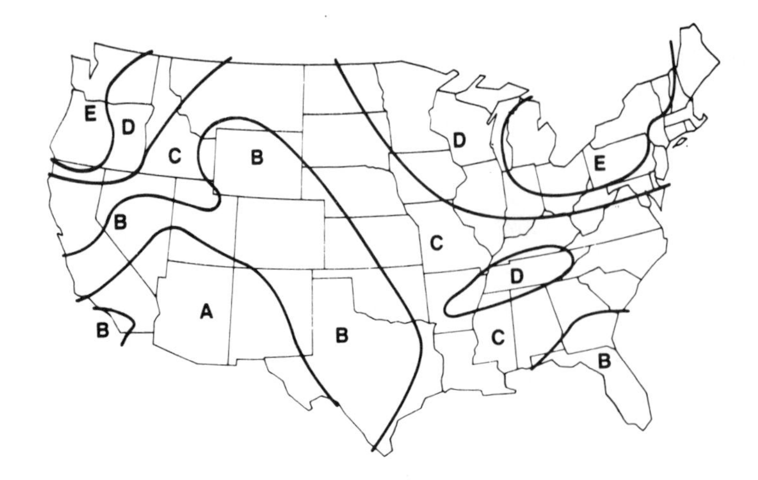


*Figure 4.11.* Location code map, continental United States.

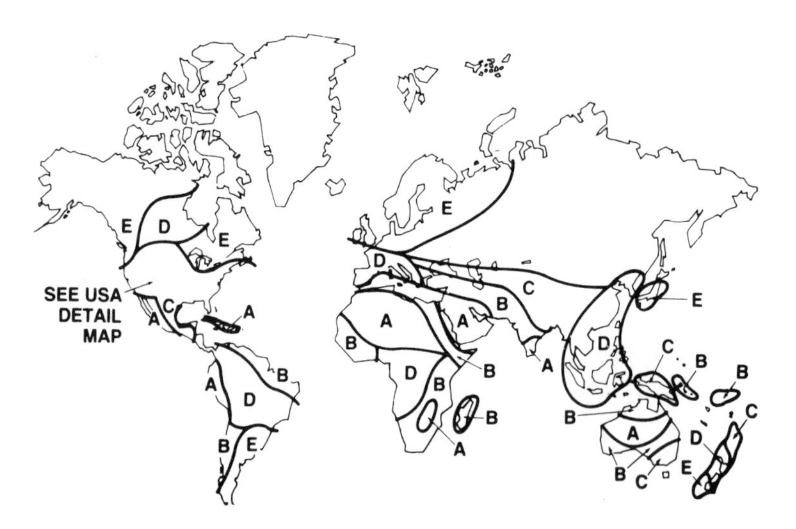


*Figure 4.12.* Location code map, worldwide.

Array output defined by this step assumes a 12-volt system and an array tilt angle equal to latitude.

### *Step 6*

Calculate battery storage capacity by multiplying load requirements from Step 1 (not Step 2) by 9.6.

*Example:*

43 AH/day x 9.6 days = 413 AH

The number 9.6 is the result of 8 “no sun” days plus a factor of 1.2. Since battery capacity may degrade 20% in 5 years, this will ensure 8 “no sun” days of capacity at the end of 5 years.

### *Step 7*

Adjust battery capacity for minimum temperature expected.

*Example:*

Assume a minimum temperature of 14°F and an available battery capacity of 78% of rated value (at 77°F). Since 413 AH (from Step 6) is needed at 14°F, then the rated battery capacity (at 77°F) would be:

Battery capacity = 413 AH/0.78 = 529 AH

## Summary

From the above procedure, it has been determined that a 43-AH/day load will require a photovoltaic array of approximately 188 watts (peak) and battery storage capacity of approximately 529 AH. The total cost of such a system, including modules, batteries, voltage regulator, wiring and steel structure, is approximately $2,500. When one considers that there are no operating costs, nearly zero maintenance, and no pollution, this system becomes very cost-effective.

While the Motorola sizing method is relatively easy to use and will specify a solar array size that will certainly meet the needs of a small, steady load, the chosen array may be larger than justifiable if an auxiliary electric system is available or if it is possible to limit electrical power usage during long cloudy spells. In that case, you may decide to try a smaller array—one-half or two-thirds the size of the calculated array—and plan to add capacity as need or budget allow. For a more detailed method of calculating the expected output of a solar array on a month-by-month basis in different parts of the country, see Rauschenbach’s *Solar Cell Array Design Handbook.* F-Chart Software and SunWatt Corporation both produce computer programs that perform sizing calculations.

The Motorola procedure calculates the peak wattage of array output needed (which is the output of the array to AM1 sunlight conditions). Some solar cell manufacturers express the output of their modules in amperes instead of watts. If they are referring to the normal 32- to 36-cell module intended to charge a 12-volt battery, it is possible to divide the number of watts needed by 16 to get the peak array current needed. (Motorola specifies the wattage of their modules as power = current x 16 volts.) So in the example above, 188 peak watts/16 = 11.75 amperes peak current.

## INSTALLING SOLAR CELL ARRAYS

Packaged solar modules are very easy to install. They are designed for outdoor use with no additional protection, so all that is needed is a rigid support that will not sway in the wind or collapse under snow. Many manufacturers sell support frames designed to hold their modules, but it is not difficult to design your own for a special application. Just remember that the array should last 25 years or more with minimum attention. Can your support structure do as well?

Large arrays can be mounted directly onto a roof or they can even be part of the roof structure. Experimental solar roof shingles have been made and, if they become cheap enough, the extra cost of the solar shingles should quickly be repaid in utility bill savings. Solar cells have been built into sailboat hatch covers and can be fiberglassed into the roof of a travel trailer.

Because of their ability to operate for years without any attention, solar cells can be permanently installed in all sorts of creative ways: onto a car roof to trickle-charge the battery whenever the car is parked in the sun; as canopies or sun shades on golf carts to extend their range; on the entire top wing of a glider to run a small prop.

### Array Orientation

In a permanent installation, it is important to orient the solar cell array so that it makes the most effective use of the available sunlight. If the array is to be fixed in place, the most useful general orientation is facing due south and tilted at an angle from the vertical equal to the latitude. If the array is to be used primarily in winter, or if that is when the short days and cloudy weather put the greatest strain on the system to meet the load requirements, the array can be tilted down a maximum of 15° more to increase its efficiency. Presumably the loss in efficiency that this would cause in summer is more than made up for by the increased amount of sunlight. If the system is used only in summer, it could be tilted up an additional 15° (maximum) from latitude. Figure 4.13 shows these three cases. The angle is not critical, however; a 15° change in angle only changes the overall efficiency by about 5%. If a roof's orientation is nearly correct, a couple of pressure-treated wood blocks might be all that is necessary to build the support.

Of course, it is also possible to change the array's orientation from time to time for more efficient operation. A simple seasonal adjustment would be no more trouble than changing your clocks to daylight savings time. Maybe in the future we will have national holidays when everybody ceremoniously readjusts their solar collectors for the season.

For concentrator arrays, some sort of adjustment is almost a necessity. Table 3.1 (Chapter 3) details the adjustments needed for different concentration ratios. The most precise adjustment would be a continuous tracking of the sun during the day. For a nonconcentrating array, this is probably not worth the mechanical complexity involved, but if the array is portable, someone could turn it occasionally to face the sun.

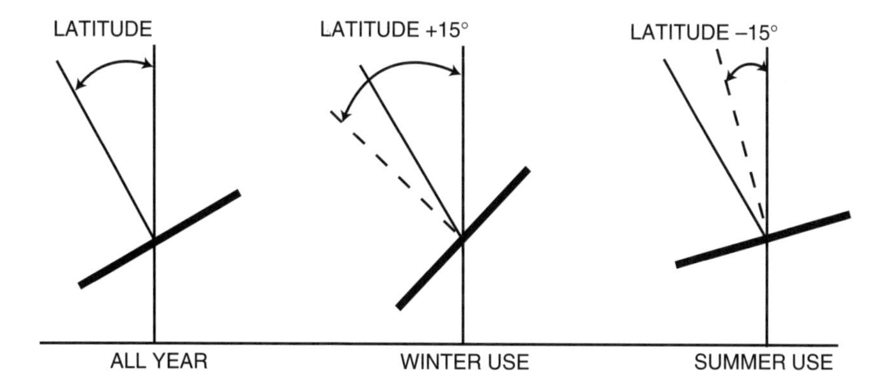


***Figure 4.13.*** Suggested tilt angles for three situations.

### Wiring the Array

The wires running from the array to the storage battery (or the device to be powered) should be carefully selected to withstand wind and weather. UV-resistant insulation and plated conductors are the best choice; the wires should be securely fastened or run in conduits so that the wind cannot whip them around and pull the connections loose from the solar array. If the array is fastened directly to a roof, it may be possible to make the connections to the back of the modules and keep the wires completely out of the weather. The wire size should be sufficient to carry the peak current produced by the array. (Appendix C lists the current-carrying capacities for different wire sizes.) A wire that is too big will not do any harm and will have more mechanical strength than the sizes given in the table for smaller currents.

Figure 4.14 shows a suggested wiring diagram for a 12-volt DC PV system. A lightning protection device would also be necessary if the arrays are to be placed in an exposed location. The blocking diode is necessary to keep the battery from discharging through the solar array at night. With a blocking diode installed, it is not necessary to disconnect the arrays when some other device is being used to charge the batteries. Some charge controllers have a blocking diode or relay built-in: check the manufacturer's instructions. A storage battery can deliver hundreds of amps into a short circuit, so a fuse should be placed as close to the battery terminal as possible to protect the rest of the wiring. This fuse should be big enough to carry the maximum expected current, but not oversized for the wire size used. The various devices that use 12-volt electricity can also be fused, of course. On a small system (less than 50 watts), you may not need a charge controller, but make sure you include fuses.

## COMBINATION SYSTEMS

A solar electric system can be combined with other alternative methods of generating electricity. A solar/wind combination is particularly good since quite often either one or the other is available. The wind system manufacturer can help you hook the storage batteries to the wind system. With a blocking diode in each system, they will not

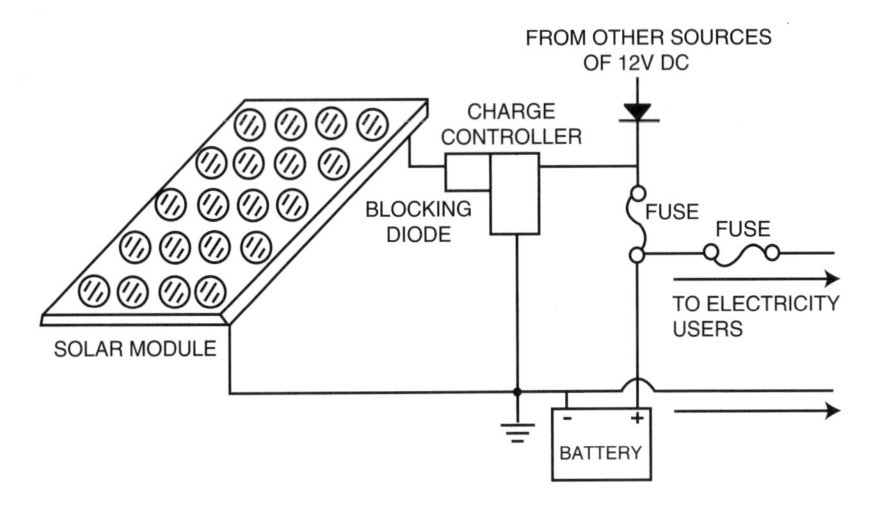


***Figure 4.14.*** Suggested wiring diagram for a 12-volt DC stand-alone PV power system.

interact with each other, but will independently charge your battery whenever they produce a sufficiently high voltage. Figure 4.15 shows a simple circuit that will power a small pilot light whenever the system is actually charging the battery. The circuit draws all its power from the charging system and uses no current from the battery.

Solar/hydroelectric, solar/generator (gasoline-or alcohol-powered) or even solar/bicycle combinations are all possible systems. The Micro-Utility System installed at the Renewable Energy Fair held in Amherst, Wisconsin, at the energy summer solstice demonstrates the simultaneous operation of a number of these systems; it even pumps electric power back into the Wisconsin Power utility grid.

## INVERTERS

All of the systems discussed so far are low-voltage DC. But because most major appliances, however, operate from 120-volt AC, the electricity produced by solar cells must be converted to AC to be usable. The device that converts direct current into alternating current is called an inverter. A great variety of these devices are available commercially. The quality of the current from the best inverters cannot

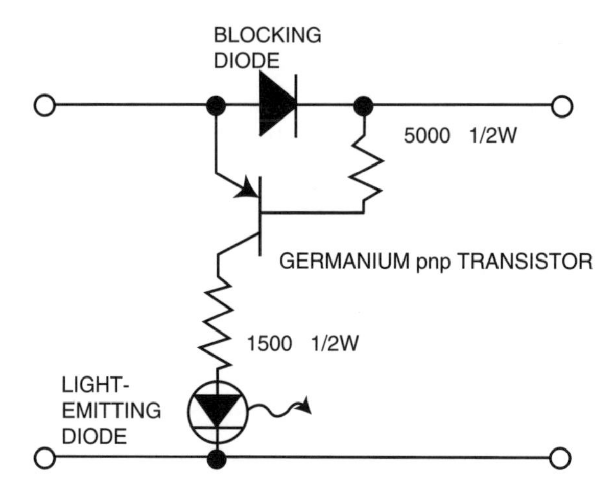


***Figure 4.15.*** A simple circuit to indicate when the battery is being charged. This circuit is powered by the solar array.

be distinguished from that of utility power in voltage, frequency, and waveform. Because of the wealth of information already available, inverters will not be covered in great depth here.

## Rotary Inverters

The rotary inverter is simply a DC motor turning an AC generator. These devices have reached a high state of development and the better ones will operate unattended for years. Although rotary inverters are quite inefficient, have moving parts, and are generally considered obsolete, older units are often available from surplus dealers for little more than the price of the scrap copper in them. Marine inverters are generally 60 cycle, but aircraft inverters are usually 400-cycle and cannot be used with most appliances.

## Solid-State Inverters

The cheaper solid-state inverters produce a modified square-wave output and large-capacity commercial models are available. For many applications (larger appliances, for example), the square waveform is not detrimental; some work with capacitors and an oscilloscope should enable the electronically minded experimenter to tune the system to produce a semblance of a sine wave. For 120-volt incandescent lights and heating elements (toasters, irons, and so on), the quality of the waveform makes no difference at all. But then, these resistance-type devices would operate just as well from 120-volt DC: an inverter won't even be needed if the solar electric system is designed to produce that voltage. For some purposes, such as record players and tape recorders, precise 60-cycle sine waves are necessary and only higher-quality inverters can be used. On a remote homestead with a minimal budget, it might be best to use DC for most applications and add a small inverter for those appliances that need high-quality AC.

The new switching inverters now made by Trace and others offer a stable sine waveform and high efficiencies at small loads. The no-load power drain of these designs is so low that the units can be left on continuously, offering the convenience of 120-volt AC electric power outlets throughout a remote solar electric home. This makes 12-volt versions of the usual home appliances unnecessary.

## Synchronous Inverters

The **synchronous inverter** is an excellent interface between a solar electric array and the available utility power. These devices, originally developed for wind energy systems but applicable to any source of DC current, convert DC to AC in synchronization with the power line. The device is simply plugged into a wall outlet and connected to the DC source. Smaller storage batteries—or sometimes none—are needed. If the solar array is producing more power than necessary at the moment, the excess is fed back into the utility grid, which then acts as the "storage" system. Utility companies are now required by law to purchase this excess power at some equitable rate, although they are understandably reluctant to participate in such a program. Synchronous inverters, commercially available from Pacific Co-Generation and others, are designed with a fail-safe relay that disconnects the inverter from the powerline if the utility power fails. This is very important because it protects utility company workers from getting shocked while repairing supposedly dead lines.

Since solar cells are still expensive, the only justification for using this synchronous inverter system is to demonstrate the techniques. But in the future, when the price of photovoltaics drops as expected, such systems may become commonplace. There will have to be a great deal of cooperation from the government and the utilities, however, to work out solutions to such problems as load leveling, networks, and large-scale storage.

## RECOMMENDED READINGS

Davidson, Joel. *The New Solar Electric Home: The Photovoltaics How-To Handbook* (Ann Arbor, MI: **aatec publications**, 1987).

F-Chart Software, 4406 Fox Bluff Road, Middleton, WI 53562.

Fowler, Jeffrey. *The Solar Electric Independent Home* [Worthington, MA: Fowler Solar Electric (PO Box 435, Worthington, MA 01098), 1991].

*Home Power Magazine*, PO Box 520, Ashland, OR 97520.

Rauschenbach, Hans S. *Solar Cell Array Design Handbook* (New York: Van Nostrand Reinhold Co., 1980).

Rosenblum, Louis, et al. "Photovoltaic Power Systems for Rural Areas of Developing Countries." *Solar Cells* 1:65–79 (1979).

Russell, Miles. "Residential Photovoltaic Power Systems for the Northwest." *15th IEEE Photovoltaic Specialists Conference Proceedings* (1984).

*Solar Today*, American Solar Energy Society, 2400 Central Avenue, G–1, Boulder, CO 80301.

Stewart, John W. *How to Make Your Own Solar Electricity* (Blue Ridge Summit, PA: TAB Books Inc., 1979).

SunWatt Corp., RFD Box 751, Addison, ME 04606.

# Chapter 5
# Batteries and Other Storage Systems

For most PV applications, some sort of storage system is needed. Many systems have been proposed or developed for the storage of electrical energy. Batteries, capacitors, flywheels, raising heavy weights, pumping water uphill, compressing air into high pressure tanks or underground reservoirs, and even converting water to hydrogen and oxygen have all been used successfully to store electricity. All of these systems are reversible, so that most of the electrical energy used to charge the system can be removed when the system is run backward. In this chapter, we will concern ourselves with the storage systems that are suitable for the average small user of photovoltaic arrays: batteries, capacitors and mechanical storage systems.

## STORAGE BATTERIES

For most solar cell applications where storage is needed **secondary** or **storage batteries** are the best alternative. A great variety of storage batteries have been developed. About the only thing some of them have in common is the ability to be recharged. (Note that primary batteries such as the carbon–zinc dry cell used in flashlights cannot be completely recharged, even once, and therefore cannot be used in any practical storage system.) All of these batteries, however, operate on the principle of changing electrical energy into chemical energy by means of a reversible chemical reaction.

Lately, a great deal of research and development has gone into new storage battery concepts, since electricity storage is the very important but weak link in solar and wind energy systems, as well as in electric vehicles. Researchers have investigated such exotic battery combinations as sodium–sulfur, lithium–metal sulfides, and aluminum–air (which isn't a true storage battery since the reaction product, aluminum oxide, must be shipped back to the refiner to be smelted back into aluminum). The most promising of these new systems is probably the lithium rechargeable battery developed for laptop computers. Large versions of the various lithium configurations are being tested for longer-range electric cars. Some of these batteries operate at room temperature while others must be heated.

We will discuss the four types of storage batteries currently available: lead–acid, nickel–cadmium, nickel–hydride, and nickel–iron.

### The Lead–Acid Battery

This battery is the most common energy storage system. The ordinary automobile battery is the lead–acid type that has evolved over the years to a compromise between low cost and reliability (for its particular use). A number of other lead–acid designs have been developed for electric vehicles, such as forklift trucks and golf carts, and as standby batteries for telephone systems and other uninterrupted power uses. A lead–acid battery designed for one use will not necessarily work well in another application, so knowledge of the different types is necessary in order to make the proper selection.

When current is drawn from the lead–acid battery, the lead oxide positive plate is converted to lead sulfate by reacting with the sulfuric acid in the battery fluid or **electrolyte**. The negative lead plate is also converted to lead sulfate. These reactions cause electrons to flow through the external circuit from one plate to another. When the battery is recharged, current is forced back into the battery, causing the reaction to reverse and restoring the plates to their original states. In a good secondary battery system, this reversible reaction can be carried out for hundreds or thousands of cycles. (See Recommended Readings for more information on the theory and operating principles of batteries.)

The size or **cell capacity** of storage batteries is expressed in **ampere hours (AH)**. This is the total amount of electricity that can be drawn from a fully charged battery until it is discharged to a specified battery voltage, and is given for a specified discharge time. For example, an automobile battery of 100-AH capacity could theoretically deliver 1 ampere for 100 hours or 100 amperes for 1 hour. In practice, the slower the discharge rate, the greater the capacity—so the capacity rating specifies the time, 20 hours in most instances (expressed as C/20). This means that if 5 amperes were drawn for 20 hours (5 amps x 20 hr = 100 AH), the battery would be completely discharged. As we will see later when battery capacity is discussed in more detail, the actual usable battery storage capacity is less than that figure. The temperature and other factors also affect the usable battery storage capacity.

The output voltage of each cell of a lead–acid battery is dependent on a number of factors—temperature and state-of-charge, for example; for the sake of categorizing batteries, it will be taken as 2 volts. The total battery voltage is the sum of the individual cell voltages, so a battery with 3 cells is called a 6-volt battery while a 12-volt battery has 6 cells in a series. Batteries are normally available in 6-, 12-, or 24-volt output, but some tractor batteries are rated at 8 volts.

First, let us consider the automobile battery, since it is the most familiar and the easiest to acquire. Automobile batteries are designed to deliver a very large current for a short time in order to operate the starter. Normally, they are not discharged more than a few percent of their capacity. The lead plates, or positive and negative electrodes, are paper-thin so that a great number can fit into a small space. This large active surface area in the plates allows currents of 200 amps or more to be drawn for a few seconds at a time without damage to the battery. Once the car is started, the alternator furnishes all the electrical requirements of the automobile and recharges the battery for the next start. If the automobile's electrical system is working properly, the battery is no longer used.

If the alternator fails and does not recharge the battery, or if you leave the headlights on overnight, the battery will become completely discharged. It can usually be recharged and will continue to work properly, but a small amount of permanent damage is done each time this happens. The normal automobile battery is designed to take twenty or so of these **deep-discharge cycles** before becoming completely useless. The battery is also designed to last anywhere from two to five years of normal automobile use, depending on the quality of the battery. This is the same as the guarantee period (five years is the longest expected life, even for "lifetime" batteries).

In newer automobile batteries the cells are sealed, with only a tiny vent hole and no caps for adding water. Two technical changes made this type of battery possible: first, a lead–calcium alloy is used in the plates to make them more stable and, second, a catalyst is placed in the top of the cell to react with the gas released during charging to convert it back into water which drips back into the cell. These "no-maintenance" batteries normally have no provision to get into the cells, but some high-quality batteries have removable caps set flush and hidden under the plastic information sheet glued to the top of the battery. Older battery cases were made of black hard rubber, but newer cases are polypropylene and some are translucent so that the fluid level in the cells can be checked visually. (The type or color of the case has no effect on battery performance.) If a set of automobile batteries is used with a solar array, a long cloudy spell could discharge them to 75% or 80% of capacity—the same as leaving the car headlights on. Because only a few such discharges would destroy the storage system, a different type of battery is required for use with solar cells.

### *The Deep-Discharge Lead–Acid Battery*

There are a number of applications, including solar cells, which require deep-discharge batteries. For example, a cruising sailboat will use small amounts of electricity over a period of days for running lights and navigation equipment before it is possible to recharge the battery. (This is one reason why solar cells are so useful on sailboats.) Electric trolling motors also need this kind of battery. A **marine battery** with its deep-discharge capability will work quite well with a small PV array.

Forklift trucks and golf carts use large-capacity deep-discharge batteries which are designed for long life and many discharge cycles. Cycle lives of 1,000 to 2,000 cycles are typical of these **motive-power batteries**, which are expected to last fifteen years or more. These batteries are heavy and expensive since they are built with thick lead plates and a greater electrolyte capacity that makes them more reliable under difficult operating conditions. A set of these batteries, if acquired at a decent price, could give years of satisfactory service at low total cost in a solar cell application.

Another common application for lead–acid batteries is in emergency standby power systems. These **stationary batteries** are kept fully charged by a regulated charging system which furnishes all the current normally needed by the system. Only in the event of a power outage is the battery actually used; the rest of the time it is "floating" in the circuit. These batteries are designed to have a very long life under these conditions (twenty-five years or more), but they have a poor deep-discharge cycle life.

Finally, manufacturers are making batteries specifically designed to be used with photovoltaic power systems. An example of such a battery is the Delco 2000 Photovoltaic Battery, but similar batteries are made by other manufacturers, such as Gould and Exide. Table 5.1 gives a summary of the characteristics of lead–acid batteries intended for different services.

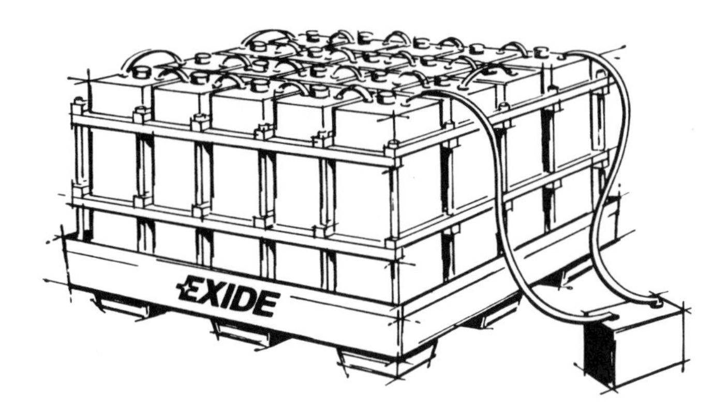


***Figure 5.1.*** Deep-cycle battery, Exide Renewable Energy Series PHv-DL. (Exide Corporation, Reading, PA)

***Table 5.1***
**Lead–Acid Battery Types**

| Type | General Characteristics | Typical Applications |
|---|---|---|
| Automotive (SLI) | High discharge rate, relatively low cost, poor cycle life | Automobile starting, lighting, and ignition; tractors, snowmobiles, and other small engine starting |
| Diesel Starting | High discharge rate, more rugged than automotive, more expensive | Large diesel engine starting; can be used for small photovoltaic systems |
| Motive Power (Traction) | Moderate discharge rate, good cycle life | Fork lifts; mine vehicles; golf carts; submarines; other electric vehicles |
| Stationary (Float) | Medium discharge rate, good life (years); some types have low self-discharge rates, poor cycle life | Telephone power supplies; uninterruptible power supplies (UPS); other standby and emergency power supply applications |
| Sealed | No maintenance, moderate rate, poor cycle life | Lanterns, portable tools, portable electronic equipment; also sealed SLI |
| Low-Rate Photovoltaic | Low maintenance, low self-discharge, special designs for high and low ambient temperatures, poor deep-cycle life | Remote, daily shallow discharge, large reserve (stand-alone) photovoltaic power systems |
| Medium-Rate Photovoltaic | Moderate discharge rate, good cycle life, low maintenance | Photovoltaic power systems with onsite backup or utility interface, requiring frequent deep-cycle operation |

### *Lead–Acid Cell Characteristics*

There are three important characteristics of lead–acid batteries to consider when designing and sizing a battery storage systems. First, the voltage output of a battery is a function of temperature and state-of-charge. Second, the useful capacity of the battery decreases significantly with a decrease in temperature. Third, batteries will slowly discharge on standing. This **self-discharge rate** is also a function of temperature and battery design.

***Output Voltage.*** Figure 5.2 shows the output voltage of a typical lead–acid cell and how it varies with the depth of discharge and the rate of discharge. At the discharge rates expected from a photovoltaic storage battery system, the curve will resemble the line marked C/10 (that is, the 10-hour capacity rate). The charging voltage curve of the battery is similar but at a slightly higher voltage. Figure 5.3 is a typical solar cell I–V curve but plotted inversely to those shown in other chapters. The two curves are remarkably similar. This similarity makes it possible, in relatively small systems, to connect the solar cell array directly to the storage battery without the need of a voltage regulator. Only a blocking diode is needed to ensure that the battery cannot discharge back through the solar cells when the array voltage is lower than the battery voltage. As the array charges the battery, the voltage of the system rises and the current output of the array drops off to that necessary to trickle-charge the battery. This system works well with 12-volt batteries if the array has 33 silicon cells in series and if the rate output of the cells is between 0.6 and 1.5 amps for every 100 AH of storage battery capacity.

This simple system will work well as long as the battery and array temperature do not vary too far from room temperature. This is true of the sailboat battery charger, for example, where the battery is inside the boat and will stay at a relatively constant temperature and the solar array will never be very cold while it is operating. If the array gets very warm, it will decrease the charging efficiency of the system but will do no harm to any of the components.

Figure 5.4 shows the effect of temperature on the output voltage of a 32-cell module and on the voltage needed to charge a nominal 12-volt storage battery. Assume there is a blocking diode in series with the array. The outputs of both solar cells and batteries drop

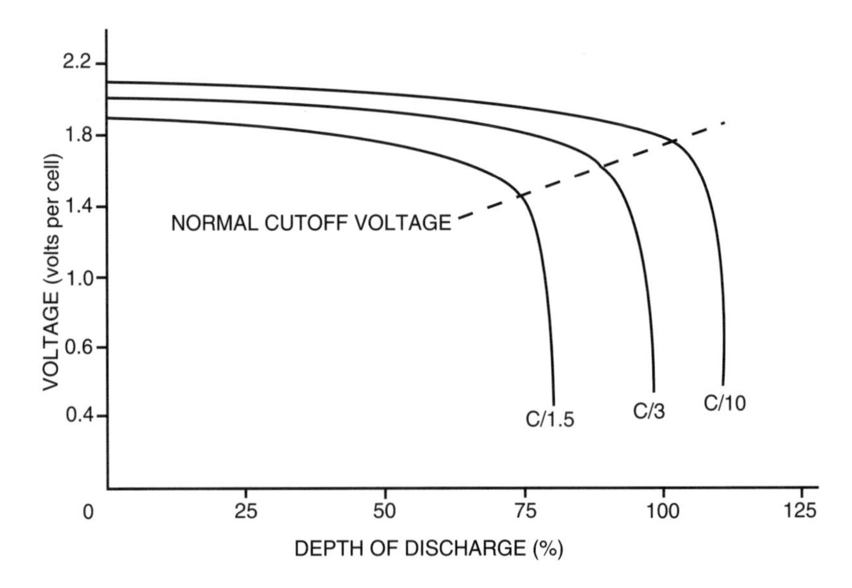


***Figure 5.2.*** Output voltage of typical lead–acid battery at different states of charge and discharge rates.

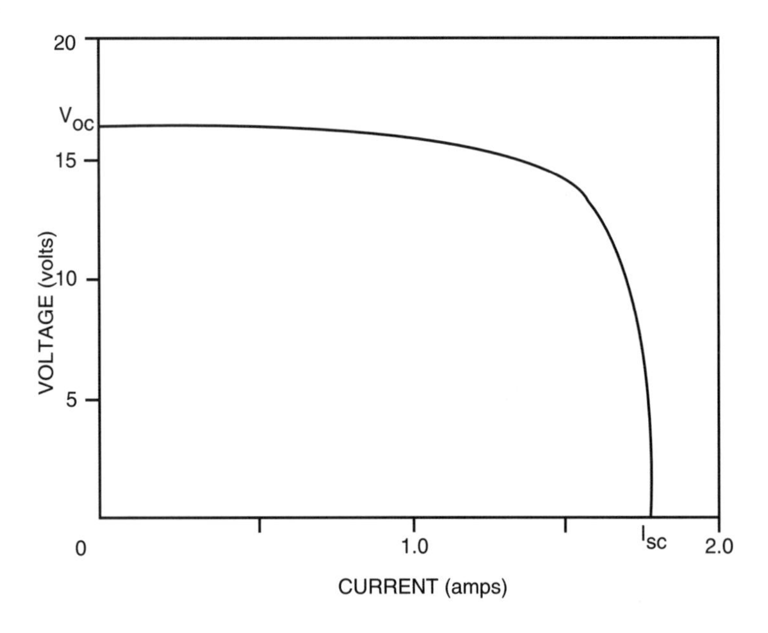


***Figure 5.3.*** Solar cell module characteristics.

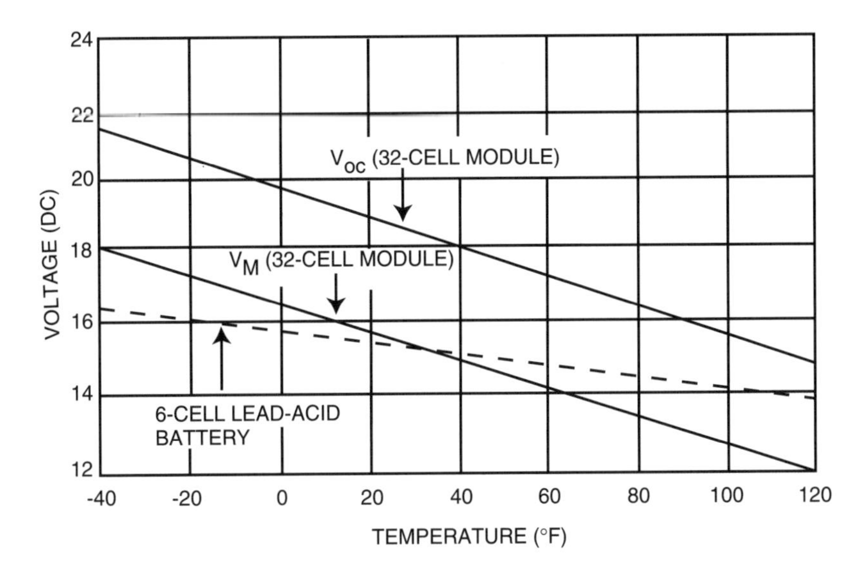


***Figure 5.4.*** Module and battery temperature characteristics.

when the temperature goes up, but they vary at different rates. At very cold temperatures it is possible that the output of the solar array could increase to the point where it is overcharging the battery, particularly if the battery is indoors or protected against temperature extremes. This overcharging will, at the least, cause gassing and excessive loss of water but it could also cause the active material of the positive plate to loosen and flake off, thereby reducing the capacity and expected life of the battery. Normally, the sun striking the solar array will heat it, minimizing the increase of output voltage with increasing sunlight intensity. In small systems where the battery capacity is relatively large compared to the amp or so of maximum output current from the array, there should be no problem. But in larger systems where the output current of the array may be large compared to the trickle-charge rate needed for the set of storage batteries, overcharging could occur under certain circumstances in cold weather and a voltage regulator might be needed. These regulators, specifically designed for use with solar arrays with complete temperature compensation built in, are commercially available. Or the simple shunt regulator described in Figure 5.5 could be built. This regulator ensures against overcharging but it does not reduce the efficiency of the PV system. If such a regulator is used, a 33- to 36-cell module might be appropriate to achieve a greater charging voltage under cloudy conditions.

***Storage Capacity.*** The amount of storage battery capacity needed is determined by the expected load and the longest period of time over which the system will be expected to operate from the batteries alone. (See "Calculating Loads" in Chapter 4 to find out how to determine the expected load.) Multiply the daily ampere-hour requirements by the number of days of storage needed. In practice, 5 to 7 days of storage is adequate to account for deviations in the weather pattern, particularly if you are at the installation and can cut back on energy consumption if the batteries get too low. For an unattended, inaccessible site where no backup is available, 8 to 10 days of storage capacity may be required and high-quality batteries with low internal leakage should be used.

Using the example of a cruising sailboat with a determined need for 30 AH of power a day—and an auxiliary engine with an alternator to recharge the system when under power—5 days of storage capacity should be more than adequate to meet the requirements: 5 days x 30 AH = 150 AH of battery capacity or two 75-AH marine

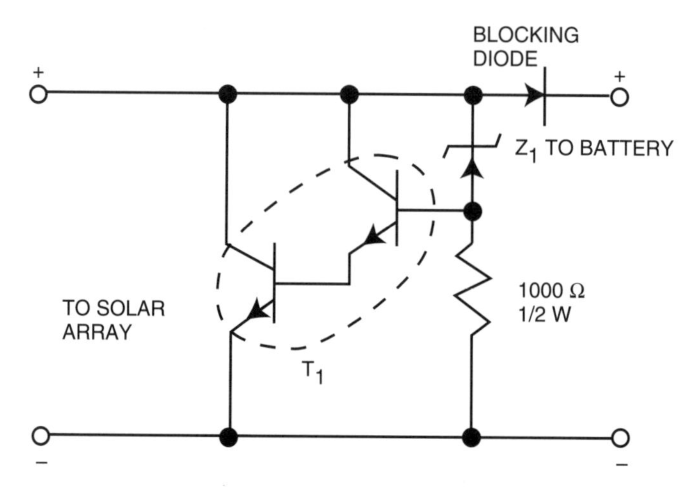


***Figure 5.5.*** Volt shunt regulator with 20 amps maximum current handling capacity $T_1$ is 2n6282 npn power Darlington. Diode $Z_1$ is IN5352 15-volt Zener diode.

batteries. It is important for safety reasons that a third separate battery be used for the auxiliary engine's starter. However, the solar pancl can be connected to this starter battery through a separate blocking diode to ensure that the battery stays fully charged even if the engine is not started for days at a time.

One important reason to have adequate battery storage capacity is that the expected life of a battery is greatly influenced by the depth of discharge in the cycles. Figure 5.6 shows this relationship for two types of batteries. In practice it is best not to discharge batteries more than 60% or so and then only on occasion. If the average discharge is less than 30%, which is what you would expect using the guidelines given above, the batteries should last their design lifetime.

The available battery capacity is a function of the discharge rate. Figure 5.7 shows this relationship for different battery types. The larger the battery storage capacity, the less the drain on the battery for a given load current. This will make the reserve proportionately larger. Doubling the number of storage batteries will more than double the capacity.

One factor that greatly reduces available storage capacity is cold weather. Figure 5.8 shows battery capacity at different temperatures. Capacity drops off sharply at temperatures below freezing, which is

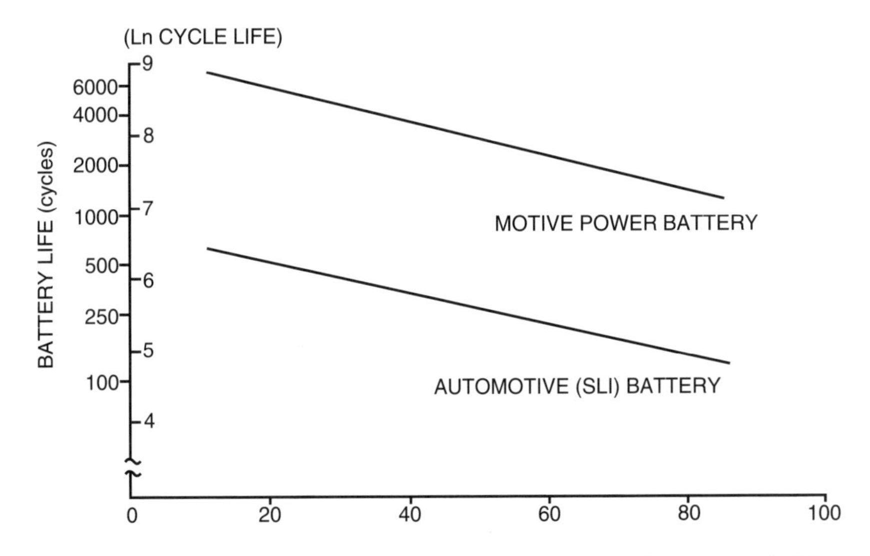


***Figure 5.6.*** Effect of depth of discharge on lead–acid battery cycle life.

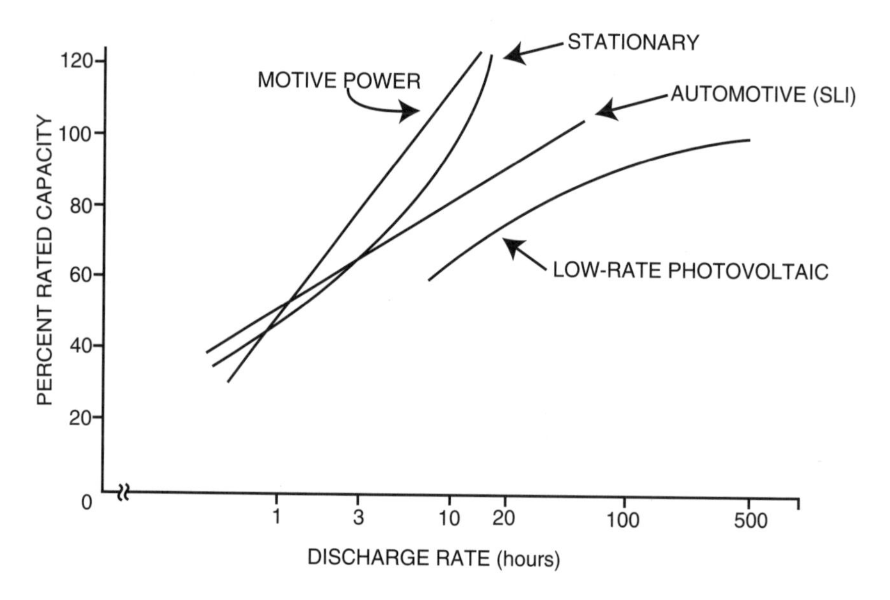


***Figure 5.7.*** Available lead–acid battery capacity as a function of discharge rate.

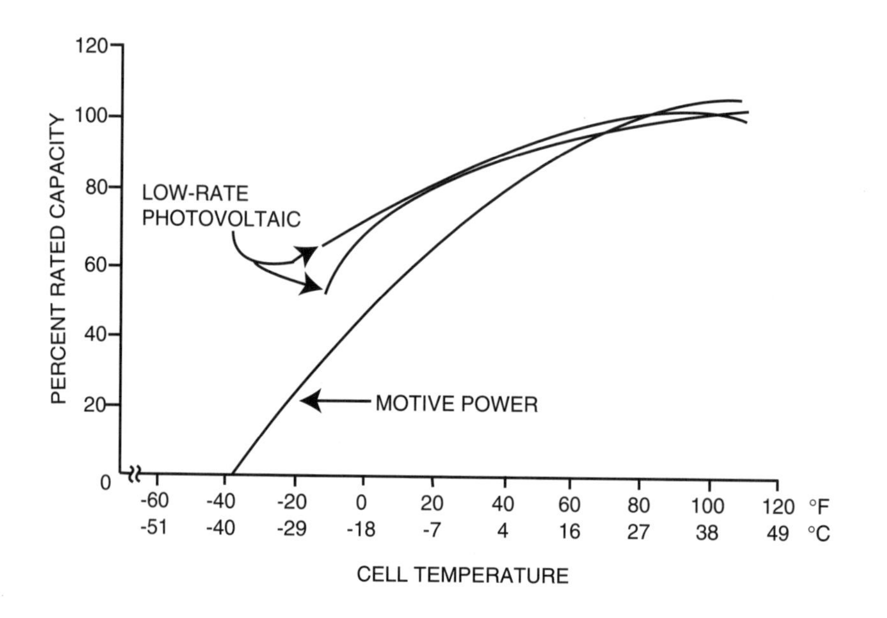


***Figure 5.8.*** Lead–acid battery capacity as a function of temperature.

one reason why cars are hard to start in the winter. Also, a battery can freeze in the winter unless it is well-charged, since the electrolyte in a discharged battery is essentially water. So it is best, when possible, to store batteries in a place that will not get too cold.

***Self-Discharge.*** Finally, batteries will slowly discharge while standing. Figure 5.9 shows how this discharge rate varies with the temperature and the composition of the plates. Most automobile batteries have lead–antimony plates, but the new low-maintenance sealed batteries normally use lead–calcium plates. As the curve clearly shows, at room temperature lead–calcium grids have a negligible self-discharge rate while the older lead–antimony grid batteries can lose their entire charge in a couple of weeks of standing. These batteries may even leak current faster than a small PV system can furnish it and the system could slowly lose charge even on a series of sunny days. Obviously, if long-term storage is needed, batteries with lead–calcium (or pure lead) plates must be used.

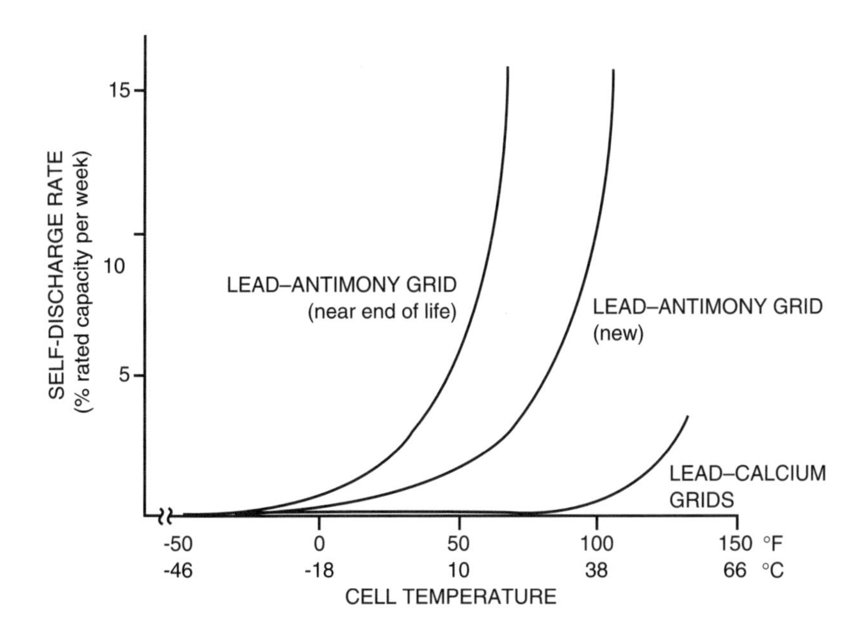


***Figure 5.9.*** Lead–acid battery self-discharge rate.

### *A Practical Battery Storage System*

From the information given above, it is clear that using lead–acid storage batteries is more complicated than placing a collection of batteries somewhere and connecting the wires.

First, the batteries must be stored in the proper environment. The best plan is to place the batteries in a rack so that they are off the floor and accessible for cleaning and maintenance. Since batteries contain sulfuric acid and give off a fine corrosive mist during charging, it is important to protect both the racks and the wiring used to connect the batteries against corrosion. The gas that bubbles up from a battery during the last stages of charging, particularly if the battery is being overcharged, is an explosive mixture of hydrogen and oxygen. The battery storage area must be adequately ventilated so that an explosive concentration cannot build up. If the battery has proper venting, a chamber inside the vents is supposed to separate the sulfuric acid electrolyte liquid from the gas, but inevitably some spray escapes and coats the top of the battery with an extremely corrosive slime. This must be washed away periodically because it is conductive and can cause leakage currents to discharge in the battery. The best way to do this is to wash the batteries first with water, then with water with a small amount of dissolved sodium bicarbonate added, and finally with water again. If the battery racks and storage area are located where a hose can be turned on them, and the floor can get wet without harm, the maintenance will be that much easier. Don't overdo the hose spraying, however, as you do not want tapwater or sodium bicarbonate inside the batteries.

The cables that interconnect the battery terminals should be heavy and the connections very tight. Automobile batteries have tapered post terminals and connectors for these are available at any auto supply store. Most other storage batteries, however, have a different style terminal with a threaded stud. This stud is usually copper- or lead-plated and so copper or lead end terminals should be used. (I have used stainless steel nuts to fasten the connectors, but copper- or lead-plated nuts also work well.) Steel nuts will corrode very quickly under these conditions. Any terminal resistance will require a higher charging voltage and waste power in the discharge cycle. The exact wiring system will depend on the battery voltage compared to the solar array voltage. This topic is covered in Chapter 1.

### *Maintenance*

Regular maintenance of the battery system is essential. In addition to cleaning, the condition of the batteries must be checked at regular intervals and water added occasionally to all but the sealed batteries. Only distilled water should be used. Regular tapwater contains small amounts of chemicals—calcium and chlorine, for example—which are bad for battery chemistry. "Rejuvenator" chemicals should not be used. If a battery cannot be rejuvenated by a proper charging sequence, it probably needs to be replaced.

A good hydrometer is necessary to check the exact charge condition of each cell of a battery. The better hydrometers have an actual float, instead of a series of balls, and a built-in thermometer with tables so the state-of-charge at different temperatures can be determined. If the battery caps are not removable and you cannot use a hydrometer, the state-of-charge can be estimated by measuring the battery output voltage when a small current (1 amp per battery is sufficient) is being drawn. Figure 5.2 can be used to estimate the condition of the battery (using the C/10 curve). Some brands of marine batteries come with an "eye" that tells the condition of at least one cell of the battery. Quite often, however, only a single cell of a battery will fail prematurely and chances are it won't be the cell with the indicator.

### *Sulfation*

The worst thing that can happen to a battery is for it to sit completely discharged and unused for any length of time. **Sulfation**—a condition where large crystals of lead sulfate grow on the plates in place of the tiny crystals normally present—will make the battery extremely difficult to recharge. Even a partially discharged battery that is not used for a period of weeks will show evidence of this. The causes and cures of this phenomenon have not been studied adequately, but sulfation seems to take place more readily at higher temperatures and may be partially reversed by a carefully controlled recharging of the battery. Start the recharging at a low current level (less than 2 amps for a 0.75-AH battery) and as the cell resistance decreases, as evidenced by a *decrease* in the charging voltage needed to maintain a constant current, increase the charging current to 10 or 15 amps until significant gassing occurs. This bubble formation will help stir the electrolyte, bringing fresh solution into contact with the plates. **REMEMBER: The gas that is given off is extremely explosive, so make sure that there is adequate ventilation and keep sparks and flames away from the battery.** If it is possible to measure the specific gravity of the battery, check all cells and continue this overcharging condition until all cells read the same. The reading should be above 1.25 (or whatever represents a full charge for the particular battery).

Some of the plate material may become dislodged and flake off during this process and the battery may never regain its original ampere-hour capacity, but at least the sulfated battery will be restored to usefulness. Very badly sulfated batteries may not respond to this treatment and will have to be traded in for their scrap value.

### *Used Batteries*

Quite often it is possible to find a set of used batteries appropriate for a PV storage system. If it is possible to purchase them for the scrap value of the lead, no money is lost if the batteries turn out to be unsuitable: you can still trade them in to recoup your investment. It is best to remember that organizations that use a large number of batteries usually have maintenance personnel who know how to care for them and generally will not get rid of them unless the batteries are pretty well finished. However, batteries that are no longer reliable for severe service (in forklift trucks or golf carts) can still last for years in a PV system if they are charged and discharged slowly and never allowed to sit for any length of time almost completely discharged.

Another source of satisfactory used batteries is dealers who maintain and replace diesel truck batteries for fleet owners. While not true deep-discharge batteries, these diesel starting batteries are built with much thicker plates and longer design lives than the best automobile batteries. When used in a system where they are discharged slowly, and rarely below 50% of capacity, they will last for a surprisingly long time. It is sometimes possible to purchase an identical set of newly removed batteries for little more than the scrap value, and since large fleet owners replace batteries on a calendar basis, regardless of use, such a set may give three or more years of satisfactory operation before they need to be replaced. Try to get a set with lead–cadmium or pure lead plates.

Stationary or float batteries that have been taken out of service by the telephone company or from standby power units may also be used. Sometimes these batteries can be disassembled for cleaning and repair. It is best to find an expert to perform this work because the procedures are complex and dangerous.

Wind energy system dealers generally know a great deal about batteries, particularly used ones, since there were once thousands of wind generators with battery storage systems on farms all over the country. One of these battery sets, even if it is 50 years old, can last another lifetime in a PV system if renovated and treated with proper care.

Used automobile batteries should be used only as a last resort or as a temporary measure until the proper battery can be found. A so-called "renovated" automobile battery is usually nothing more than a traded-in battery that has been cleaned, repainted, and charged. Since many people buy a new battery when the real problem with their autos is some other electrical component, a traded-in battery may have a lot of life left.

### Nickel–Cadmium Batteries

First developed after the turn of the century, nickel–cadmium batteries were not commonly used in the United States until the 1950s. The two plates of this battery are a nickel plate packed with nickel oxide and a cadmium plate. The main advantages of the nickel–cadmium (ni–cad) battery are long life and reduced maintenance requirements. The main disadvantage is the high cost per ampere hour of capacity. In smaller systems, the convenience of a sealed battery semipermanently installed in a device outweighs this higher cost. Most rechargeable flashlights, power tools, and pocket calculators use some sort of nickel–cadmium battery.

Ni–cad batteries are manufactured in two types: "sealed" and vented. The sealed type has a pressure relief valve built into the cell to prevent an explosion if the cell is heated or greatly overcharged. Some of these valves will reclose after they pop open; others stay open allowing the water to evaporate from the electrolyte, quickly ruining the battery. The vented type have resealable vents that open and close under small pressure changes. The electrolyte, a solution of potassium hydroxide in water, is used only to carry the current between the plates and does not change when the cell is charged or discharged. The nickel–oxide electrode is either of sintered-plate or of pocket-plate construction; the sintered plate is the one most used in small cells of the sealed design.

The voltage output of a ni–cad battery is 1.2 volts per cell and changes very little with use until the battery is almost completely discharged, at which point it will drop sharply. A cutoff voltage of 1.0 volt is usually used as an indication that the battery is discharged. A ni–cad battery can accept a charge at a relatively high rate (C/1) and is capable of operating under continuous overcharge provided that the charging current does not exceed a given amount (C/15). To give a numerical example, a nickel–cadmium sealed cell with an ampere-hour capacity of 15 could initially stand a charging current of up to 15 amps, but as the cell becomes charged, this would have to be reduced to a final current of 1 amp. This 1-amp current could run through the battery indefinitely without harming it. Of course, the battery could also have been charged at a steady rate of 1 amp from the beginning.

Nickel–cadmium batteries can be deep-discharged many times without damage and have a much smaller change in performance with temperature compared to the lead–acid battery. All these properties make them ideal storage batteries to use with PV systems where the battery bank can be connected directly to the solar array with no need for a voltage regulator. (A blocking diode would still be needed, however, to keep the array from discharging the batteries in the dark.) It is important that the maximum output current of the array not exceed the C/15 continuous overcharge current of the battery.

Ni–cad batteries have one quirk that needs mention: this is the "memory" effect. For example, after a ni–cad battery has been repeatedly discharged by 25%, the cell voltage drops off sharply at this state-of-discharge and the battery acts as if it has only one-quarter of its true capacity. This effect is shown in Figure 5.10 together with the voltage versus discharge characteristics of a normal cell. Prolonged periods of overcharge will also produce such an effect. In most cases, the battery can be restored to full capacity by letting it discharge until it is completely dead and then recharging it completely, allowing it to overcharge to a voltage of 1.7 volts per cell. Since this can

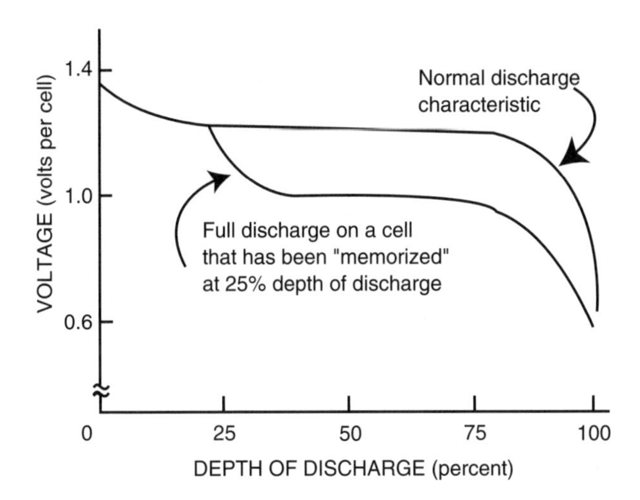


***Figure 5.10.*** Memory effect on discharge voltage of sintered plate nickel–cadmium battery.

equal 17.5 volts for a typical 10-cell nominal 12-volt system, 12-volt loads and appliances have to be disconnected during this equalization charge period. This complete discharge cycle should be performed about once a year on a regular maintenance schedule for large ni–cad battery storage systems.

In conclusion, ni–cad batteries are very useful for small solar-powered systems but they are too costly for larger systems unless the increased reliability and low maintenance requirements are worth the premium price. This is the case for critical equipment at a remote or inaccessible site, such as a Coast Guard buoy or a space satellite power system.

## Nickel–Iron Batteries

These batteries were developed by Thomas Edison in the late nineteenth century as a more reliable alternative to the lead–acid battery and were used extensively by the railroads as storage batteries on passenger cars. They are extremely rugged devices, not bothered by the abuse that would completely destroy the usual lead–acid battery. However, they do have some electrical disadvantages. The first, not greatly important for stationary use, is that the output voltage is low—

about 1.1 volt per cell—so that approximately twice as many cells are needed to produce the same battery voltage. Coupled with the relatively large plate size needed for a given storage capacity, this produces a much larger and heavier battery for the same ampere-hour capacity. The second disadvantage is that the output voltage drops more quickly as the battery is discharged so that a half-discharged nickel–iron battery has a much lower voltage than a half-discharged lead–acid battery. If the battery were used in a float system where under normal operation the state-of-charge changes slowly and over a small range, the change in output voltage would be minimized. Another solution is to design the photovoltaic and storage system to produce an excess output voltage when fully charged and use a voltage regulator to control the final output voltage.

A more practical problem with Edison batteries is finding them. Eagle Pritcher and SAFT are just starting to manufacture a new style of nickel–iron battery as a replacement for forklift and other industrial lead–acid batteries. Although these newly designed batteries have a good storage capacity for their weight and some of the electrical discharge problems have been eliminated, they are currently very expensive and not readily available. The original Edison-style battery is currently not being manufactured, although surplus dealers may have old units removed from passenger cars. If you can find a set of these batteries in good operating condition and can locate someone who is familiar with their care and maintenance, you should end up with a first-class electrical storage system at a very low cost.

## Nickel–Hydride Batteries

This newly developed battery uses metal hydrides as one of the electrodes and is really a form of self-storing hydrogen electric system. The metal–hydride cell has some real advantages over the nickel–cadmium battery it closely resembles. First, it has a greater storage capacity for a cell of the same size and weight. Second, it seems to have no "memory effect," so it requires less maintenance and less careful charging cycles. Third, it contains no environmentally destructive cadmium (although other metals or compounds in the battery could be a hazard). These premium batteries are available in the AA, C, and D sizes to replace ni–cads, and are being used in most of the newer laptop computers to give a longer operating time between charges. They are expensive, about twice the price of ni–cads but without twice the storage capacity. The nickel–hydride battery is just now being produced in larger sizes suitable for electric vehicle and stationery storage systems, but the price and lack of easy availability will keep them from general use in PV systems for several years. However, since people are setting electric vehicle range records using these batteries, I suspect they will be the next generation storage battery if the price comes down.

## SMALL STORAGE SYSTEMS

For digital watches and pocket calculators, small, so-called rechargeable batteries are sometimes used. In these micropower systems, the distinction between primary and rechargeable batteries becomes lost as, at a very slow rate, any battery action is reversible to an extent. Many "solar" digital watches use the same silver oxide or mercury battery that their nonsolar counterparts use; they rely on the solar cells to extend the normal lifetime of the batteries, which is measured in years in any event. Work on a truly rechargeable tiny battery is now under way in Switzerland and Japan, as well as in the United States. The most interesting concept is a solid ceramic battery in which diffusion of small ions through a crystal or amorphous matrix replaces the liquid electrolyte. Such batteries should have useful lifetimes (in watches, for example) of decades. Similar batteries using lithium ions diffusing through porous plastic sheets are starting to be used in expensive laptop computers.

For very small loads, a capacitor may serve to store small amounts of current for a short time and to smooth short-term changes in the load or solar cell output. For example, a 4000- to 8000-microfarad electrolytic capacitor can aid in delivering the starting current for a small motor when the running current is furnished entirely by solar cells. A capacitor can also be used to ensure a continuous supply of power to a solar-powered calculator when there is no battery or on/off switch. This will prevent a momentary shadow from shutting off the integrated circuit and erasing the calculation in progress. Recently, very large **ultracapacitors** have become available. These experimental carbon-based electrolytics (with capacities measured in farads instead of microfarads) can be used for delivery of very large currents for accelerating big motors; they have even been proposed as replacements for batteries in electric vehicles.

Another very experimental energy storage system uses superconducting rings to store energy in intense magnetic fields. Experimental prototypes were built as part of the U.S. government's "Star Wars" program. These expensive devices needed to be cooled with liquid helium but because it takes a good deal of energy to liquefy helium, their total efficiency was low. However, the development of higher temperature copper–oxide superconductors makes a simpler device theoretically possible. These **superconductor storage rings** would give an electric car a range of 1,000 miles or more. Perhaps sometime in the twenty-first century, we will be driving solar electric cars with seasonal storage capacity averaged over months of operation.

## MECHANICAL STORAGE SYSTEMS

Most mechanical storage systems are either too complicated or are applicable only to large-scale use. Nonetheless, there are situations where mechanical storage can be very useful, even in small systems (for example, where the electricity is just a medium and the desired end product is stored). One example is a water pump that runs while the sun shines, pumping water into a reservoir to be drawn upon as needed. Another example is a solar-powered refrigerator that has holding plates that can keep food cold for two or three days when completely charged by the photovoltaic-powered electric compressor. This idea could be extended to air compressors with large storage tanks and similar systems. A simple diode or relay switching system could be used to direct the solar array output to a set of storage batteries when the system is charged or filled to capacity.

## FLYWHEELS

Flywheels have been proposed as small- and medium-sized storage systems and have actually been used in some European homes for years. Considerable development needs to be done, however, before they will be available in easy-to-use, packaged systems. With the renewed interest in electric vehicles, several groups are developing experimental flywheel systems using space-age materials like boron fiber composites. The prototype systems, built under contracts with the Applied Research Projects Agency (ARPA) of the Department of Defense, are able to efficiently store four to six times the energy as the equivalent weight in currently available storage batteries. The units built so far seem to have a long service life: they run in a vacuum housing and use magnetic bearings so that there are no moving parts in contact with one another. The fiber composite flywheels (counter-rotating to cancel the gyroscopic effect) safely disintegrate to fluff if they come apart so that no shrapnel will escape the housing, but questions of long-term performance and the ability to withstand the bounce of hitting potholes still remains to be answered for vehicular use. Flywheels could replace battery storage systems in the near future, however, in large stationary installations, such as power company substations. They would be an ideal complement to neighborhood-sized photovoltaic arrays.

## HYDROGEN FUEL CELLS

It is very simple to use the low-voltage DC output from a series string of three or four solar cells to separate water into hydrogen and oxygen. Put a pair of electrodes into the water and connect them to the solar array. A few drops of sulfuric acid in the water will make it sufficiently conductive to pass the necessary current. This system will utilize 10 to 12% of the energy in sunlight to make hydrogen. This hydrogen can efficiently be used to recover the electricity in a hydrogen fuel cell.

It is also possible to produce hydrogen directly in a photoelectrochemical cell, but the efficiency of the best of the experimental systems developed so far falls short of the simple indirect system described above. This is because, in the indirect system, the materials used in the light absorption and hydrogen evolution steps can be optimized for their particular functions.

There has been a great deal written about the **hydrogen economy**, a system in which hydrogen is substituted for fossil fuels in many technological applications in transportation and industry. However, there is one important point to remember: hydrogen is not a primary energy source; it is a way of storing energy. All the energy that can be generated by using hydrogen represents energy that was put into the hydrogen when it was electrolyzed from water. Because

***Table 5.2***
**Comparison of Battery Types and Flywheels**

| | Lead–Acid Deep-Cycle | Nickel–Cadmium | Nickel–Iron | Nickel–Hydride | Nickel–Zinc | Lithium–Polymer | Flywheel Composite |
|---|---|---|---|---|---|---|---|
| **Volts per Cell** | 2.1 | 1.2 | 1.1 | 1.2 | 1.2 | 3.0 | N/A |
| **Energy Density (Whr/kg)** | 35 | 50 | 60 | 60 | 70 | 80 | 150 |
| **Life(cycles)** | 500 | 2,000 | 500+ | 1,000 | 300 | 1,000 | 5,000+ |
| **Calendar Life (years)** | 5–8 | 10–15 | 10+ | 8+ | 3 | unknown | 15+ |
| **Cost ($/kWh)** | 200 | 1,500 | 2,000 | 2,500 | unknown | 3,000+ | unknown |

of the inefficiencies of the process, more energy—from some other source—is used to make the hydrogen than can be extracted later from the hydrogen.

The energy stored in hydrogen can be used in a number of ways. Hydrogen can be burned like natural gas to furnish heat, it can be used as a fuel in a piston engine or turbine, it can be oxidized in a fuel cell to generate electricity, or it can even be burned in a rocket engine to power a space shuttle. Thus, using sunlight to produce hydrogen leads to a very flexible set of options. However, hydrogen made in this fashion is expensive and represents a good deal of high-quality electrical energy. It would be much cheaper and more efficient, for example, to use a solar water heater and store the heated water than to produce hydrogen with a photovoltaic system and later burn the gas to heat water.

The hydrogen economy discussed by some authors is conceived of as a large-scale network of hydrogen producers, pipelines, and users similar to our present natural gas utility system with the important difference being that hydrogen is a renewable source of energy whereas natural gas is a fossil fuel in limited supply. Presently, most natural gas is used to produce heat that could be furnished directly by the sun through relatively simple systems. The hydrogen economy would make the most impact in the future in systems that need more logistics to transport and store electricity than hydrogen.

Hydrogen is also a very clean-burning fuel that could be used in automobiles or other transportation systems. However, it is a very light gas and would have to be condensed in volume before a vehicle could carry sufficient fuel. The energy required to compress a volume of hydrogen to the 300 pounds or so necessary to fill a tank of reasonable size so that it can power the vehicle for 200 miles or more, or the energy needed to cool the gas and produce liquid hydrogen (which also occupies a much smaller storage volume) is an appreciable fraction of the energy that will be released when the hydrogen is burned. Of course, some of this energy might be recovered in an expansion engine on the vehicle. However, the best way to store hydrogen in a small volume is to react it chemically to form a liquid solid product. Metal hydride systems are being developed that soak up and release large quantities of hydrogen with small pressure changes at room temperature. Calculations indicate that such systems could carry enough hydrogen in a small volume to make them practical for transportation systems. The best way to utilize hydrogen in a vehicle is to run a hydrogen fuel cell to produce electricity for wheel-mounted electric motors. Besides being very flexible in operation, the overall system is much more efficient than an internal combustion engine and creates no combustion products, such as nitrogen oxides, to pollute the atmosphere.

Hydrogen can also be reacted with carbon dioxide from the air to make solid or liquid organic compounds such as alcohols. These can be burned as fuel or used as starting materials for a wide variety of products including food, drugs, and plastics. The best-known and most completely developed such system is photosynthesis by plants. Here the energy of sunlight is used to separate hydrogen from water and combine it with carbon dioxide to form sugars. Later, the sugars are rearranged and assembled into an array of substances. Carbohydrates or sugars represent a very convenient way of storing the energy from sunlight. In a sense, all life depends on this hydrogen economy.

## RECOMMENDED READINGS

American Flywheel Systems, Inc., PO Box 449, Medina, WA 98039. Information packet.

"Battery Service Manual." The Battery Council International, 401 North Michigan Avenue, Chicago, IL 60611.

Bockris, J. O'M. *Energy. The Solar–Hydrogen Alternative* (New York: John Wiley & Sons, 1975).

"Capturing the Sun—Batteries and Solar Energy Storage." Lead Industries Assn., Inc., 295 Madison Avenue, New York, NY 10017.

*Handbook for Battery Energy Storage in Photovoltaic Power Systems* (San Francisco: Bechtel National, Inc., Research and Engineering Operation, 1980).

Vinal, George. *Storage Batteries*, 4th ed. (New York: Wiley lnterscience, 1955).

# Chapter 6
# New Developments in Photovoltaic Technology

The standard method of making silicon solar cells is both expensive and energy-consuming. If solar cells are going to compete with the nonrenewable resources now used to generate electricity, new types of cells and new manufacturing processes will have to be developed. The present price of about $4 to $5 per watt for crystalline silicon modules is too high to compete with central utilities. The Department of Energy has an ongoing research program aimed at reducing this cost and this program, together with the photovoltaic part of the Small Business Innovation Research (SBIR) Program, is succeeding in lowering cost as well as increasing module longevity. In addition, a number of private companies and research groups are still working on a great many alternative approaches to photovoltaic production of electricity. These include new ways of using silicon, other materials and combinations for solar cells, and clever optical systems to more effectively exploit the photovoltaic process. This chapter will review a number of these systems and will attempt to assess their prospects.

## OTHER JUNCTION STRUCTURES

Many of the new developments start with silicon, the best understood semiconductor material. Once the thin flat piece of silicon is produced, a junction has to be formed at the top surface to make a photovoltaic cell. A number of alternatives to the diffused p–n junction are being investigated, both to increase cell efficiency and to cut manufacturing costs.

Martin Green and his co-workers at the University of New South Wales in Australia have been working on entirely new configurations for the PV junction in silicon. Their latest work uses thin layers of alternating p- and n-type silicon deposited onto glass substrates. Vertical grooves are etched by laser beam through these layers all the way to the glass substrate and then metal contacts are added to fill the grooves. By making alternating buried metal strip contacts either p- or n-type silicon, a large number of tiny p–n junctions are created in one cell. With its extremely short minority-carrier path lengths, this configuration is very tolerant of defects and acts almost like the theoretically ideal silicon solar cell. The grooved, clear top coating bends light that would have fallen onto the metal fingers away to the active cell area, utilizing almost every photon and producing a record silicon-cell efficiency of over 24%. The process is licensed to BP Solar, which has simpler buried-contact configuration modules already in production in Australia.

### Tin Oxide and Indium Oxide

Two successful heterojunctions with p-type silicon are **tin oxide** and **indium oxide** doped with tin oxide (called **ITO** by the engineers working on the material). Both are n-type semiconductor materials which are transparent to visible light but which are good conductors of electricity. Tin oxide can be easily deposited onto glass or other substrates by spraying tin chloride in water onto the heated substrate. Tin oxide–coated glass (called NESA glass) is used as a transparent conductive surface for such applications as aircraft windows. The processes for producing it are well-developed and inexpensive. Indium oxide thin films, a more recent development, are easy to produce and are used in the "Low E" insulating glass used in modern house windows. However, indium is an extremely rare element and there are those who question the advisability of making solar cells with indium compounds, since such a large quantity of cells would be required to meet a significant portion of our energy needs. In the laboratory, tin oxide and ITO heterojunctions with silicon have been produced which are as efficient as the normal p–n homojunction silicon cells.

### Metal–Semiconductor Junctions

Another type of junction that can be used in a solar cell is the **Schottky barrier** or **metal-to-semiconductor junction (MS junction)**. This junction is produced when a low work-function metal like aluminum is deposited onto a p-type semiconductor or a high work-function metal like gold is deposited onto an n-type semiconductor. Becquerel's first photovoltaic cell was a Schottky barrier device. This barrier contact is in contrast to the low-resistance ohmic contacts usually required for metallic contacts to semiconductor devices. If an extremely thin (20 Å) insulating oxide layer is placed between the metal and the semiconductor, the device is called a **metal–insulator–semiconductor junction** or **MIS solar cell**. MS and MIS junctions have been made with silicon and nearly all the metals in the periodic table. In the laboratory, some of these devices have shown enhanced open-circuit voltages and efficiencies compared to the standard silicon p–n homojunction solar cell; however, the complexities involved in making the precisely controlled, extremely thin layers required for their successful operation would seem to rule them out as likely candidates for commercial devices in the near future. The research devoted to these junctions has greatly increased our knowledge of how semiconductor devices work and of some of the mechanisms that cut solar cell efficiency.

## PHOTOELECTROCHEMICAL CELLS

A few years ago, Texas Instruments (TI) announced the development of a new type of silicon solar cell. Instead of an all solid-state device with the junction between two different solid materials, this cell has a silicon–liquid junction. The silicon electrode is not in the form of a flat surface at all, but rather tiny beads made in the same way as buckshot. Molten silicon is poured through screens to make tiny droplets that solidify as they fall through the "shot tower." The simplified processing steps and the ability of the device to produce hydrogen as well as electricity had created quite a bit of excitement, given the interest in the **hydrogen economy**. More recently, Texas Instruments has developed a method of mounting the beads between a perforated aluminum sheet and a solid backing sheet to produce an inexpensive, flexible silicon solar cell. TI has joined with Southern California Edison to set up a manufacturing plant to exploit this process to produce PV modules for large-scale solar power farms.

The **liquid junction** photovoltaic cell is also being investigated in a number of laboratories. Most of these cells use a semiconductor material as the anode and have had problems with long-term stability, because the light-induced chemical reactions can irreversibly change the anode surface. But Bell Labs has developed a type of liquid junction cell that has the active semiconductor electrode as the cathode and the cell actually becomes more stable when exposed to sunlight. The device can be designed to produce either electricity or hydrogen, and steady-state efficiencies of 11.5% have been achieved. However, their current cell uses scarce indium phosphide as the photocathode material; any large-scale exploitation of this technique will necessitate the development of a more commonly available cathode material.

## CADMIUM SULFIDE/CADMIUM TELLURIDE AND OTHER II–VI HETEROJUNCTION CELLS

In the 1960s, Clevite Corporation developed a thin-film solar cell to be used in the space program. The cadmium sulfide solar cell, as it was called, had been invented in the early 1950s by Reynolds. It was not until twenty years later, however, that scientists figured out that the active semiconductor in the cell is not cadmium sulfide, but rather the cuprous sulfide layer found when the cadmium sulfide layer is dipped into a copper chloride solution. The cell is another example of a heterojunction solar cell.

The Clevite cell was never very successful because of difficulties in reproducibility of good cells and long-term instabilities in the finished modules. In the early 1970s when interest in terrestrial uses of solar energy made a low-cost solar cell attractive, major research on the CdS/$Cu_2S$ cell was again undertaken, mainly at the University of Delaware under Karl Böer. For several years, SES, a subsidiary of Shell Oil, sold a solar module that used vacuum-deposited CdS, but it was taken off the market in 1976 due to instability problems. Because this type of cell can react with the moisture in the air, it was found that traces of moist air entered the modules—even when sealed into glass-fronted enclosures—and destroyed the junctions. SES tried to market the devices several times but found the sealed modules required were too expensive to market competitively.

Ease of manufacture from relatively inexpensive starting materials has kept interest alive in this potentially low-cost solar cell. J. F. Jordan of the Baldwin Piano Company developed an inexpensive way to spray cadmium sulfide coatings onto glass substrates. His process was taken to the pilot-plant stage by Photon Power Corporation of El Paso, Texas, a joint venture between Total, a French oil company, and Libbey–Owens–Ford. Photon Power had expected to sell 4% to 5% efficiency modules at $2 per watt of capacity when they went into full-scale production, but start-up problems caused them to drop the project.

Although the cadmium sulfide/cuprous sulfide solar cell turned out to be unstable and not commercially usable, cadmium sulfide is still an important photovoltaic material. Its best use at present is as the window material in heterojunction cells that have replaced the thin cuprous sulfide layer (that was the actual photovoltaically active material in the old "cadmium sulfide" solar cell) with other materials. There are a number of materials similar to cuprous sulfide that can be used to produce photovoltaic junctions. These II–VI compounds (named after the second and sixth columns of the periodic table) are being investigated and several promising materials such as cadmium selenide and cadmium telluride have been used in solar cells. These materials can be deposited as thin films by essentially the same processes that are used for cadmium sulfide films and also by electroplating. Ametek, Incorporated of Harleysville, Pennsylvania, and BP Solar in England have both electroplated N–I–P cadmium telluride solar cells, with BP's cells reaching 13% efficiency. Other processes such as silkscreen printing and spray coating have produced thin film CdTe cells of similar efficiencies, but new processes such as atomic layer epitaxy using close-spaced sublimation have successful-

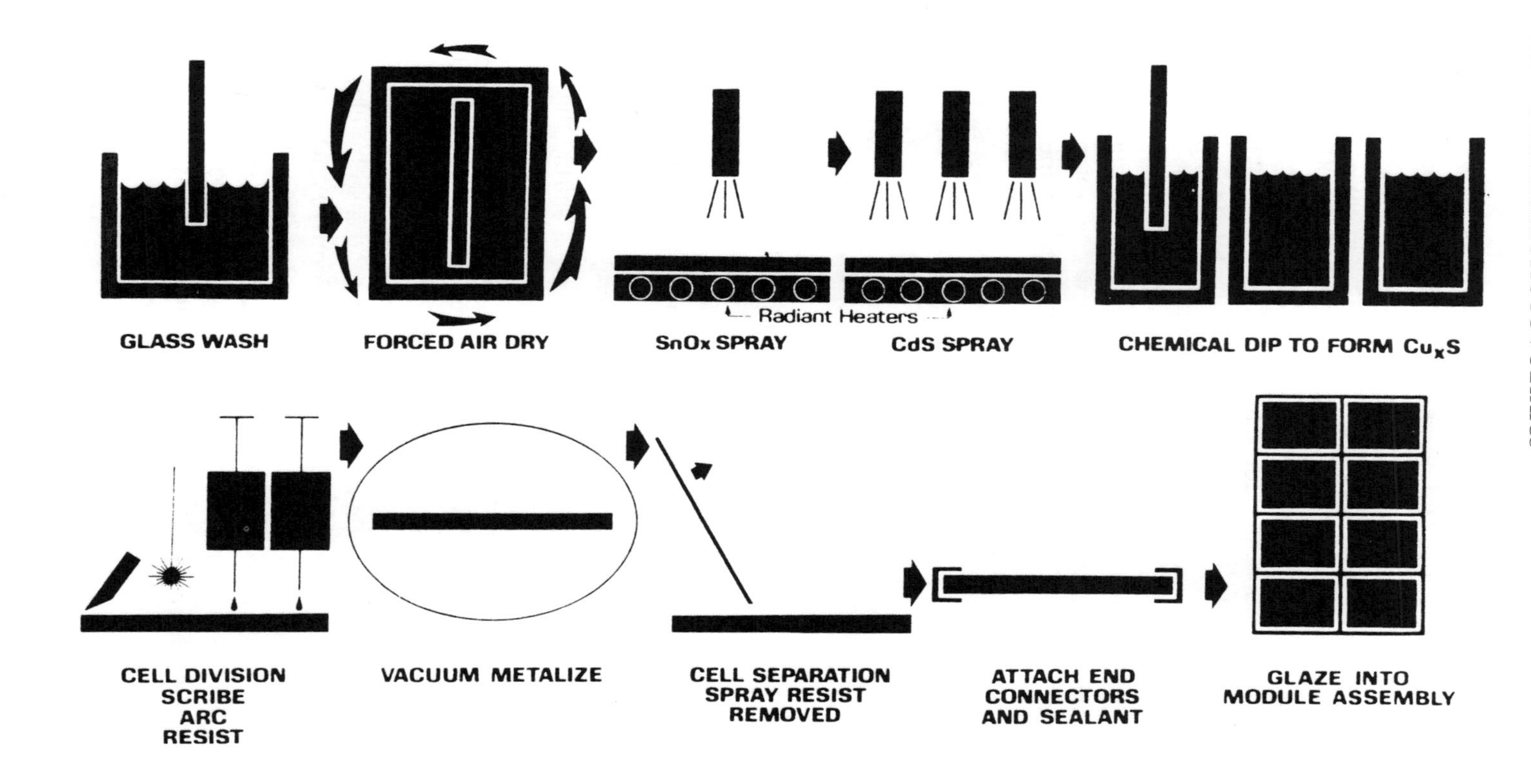


***Figure 6.1.*** Process of producing thin-film photovoltaic cells. (Photon Power, Inc., El Paso, TX)

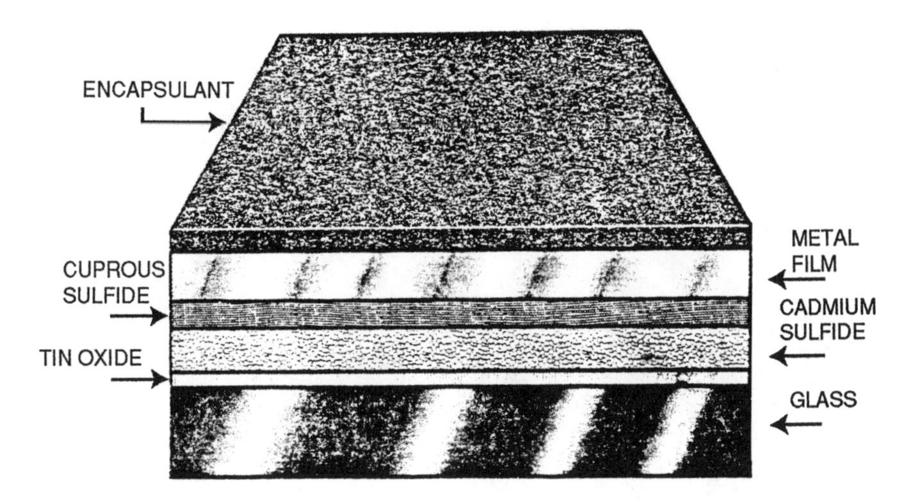


***Figure 6.2.*** Diagram of a cadmium sulfide/cuprous sulfide cell. (Photon Power, Inc., El Paso, TX)

ly produced CdS/CdTe cells with a verified efficiency of almost 16%, a new record for this material. Boeing Corporation and others have replaced the cuprous sulfide layer with one of copper indium disulfide and have achieved a thin-film cell with over 13% efficiency. A great number of combinations are possible and the development of these semiconductor materials is now the focus of a great deal of research at the National Energy Research Labs (NERL) and a number of semiconductor manufacturers. In fact, Matsushita in Japan is now delivering small silk-screened cadmium telluride modules for use in pocket calculators; manufacturers in Europe will soon follow suit.

Some questions have been raised about the safety of solar cells containing cadmium because it is a poisonous heavy metal. Of course it is extremely important that introduction of a new technology not be a cause of new environmental problems when applied on a large scale. It is ironic, however, to compare the accessibility of cadmium in solar cells to that used to plate refrigerator racks and the bolts on children's toys. Recently, the DOE has begun to research methods of recycling thin-film PV modules containing cadmium to recover the heavy metals, both to protect the environment and to reuse the relatively scarce cadmium, tellurium, and indium.

***Figure 6.3.*** Town of Paradise Valley, Arizona, police repeater on Mummy Mountain. (Solavolt International, Phoenix, AZ)

## III–V SEMICONDUCTOR COMBINATIONS

The III–V group, a whole class of compound semiconductor materials, derives its name from the third and fifth columns of the periodic table. The compound semiconductors are made by choosing one (or more) element from each column. Gallium arsenide is the best known of these substances, but the list also includes gallium phosphide, indium phosphide, and more complex mixtures such as aluminum gallium arsenide. The III–V compounds are important commercially; for example, they are used in making those light-emitting diodes (LEDs) lighting up all over new appliances.

The III–V semiconductors have been studied extensively; their properties are well understood and they can be tailored to specific applications. The band gap of gallium arsenide is almost ideal for the construction of high-efficiency solar cells; the high-quality crystals made of these materials allow most of the light-generated current to be collected by the contacts. The material will also withstand surprisingly high temperatures without damage. Consequently, the ~28% efficiency attainable with these solar cells represents the greatest efficiency of any solar-cell single junction produced so far. Several groups are working on extremely high-quality III–V compound solar cells for use at the focus of high-power concentrating systems. Boeing has achieved a record 33% efficiency with a multijunction gallium arsenide/gallium antimonide concentrator cell. The only problems with these cells are cost and availability.

All the III–V compound solar cells developed so far use gallium and/or indium. Both elements are extremely rare, much more so than gold, and the only reason the prices of these metals are not astronomical is that until recently there was essentially no commercial use for them. They are produced as by-products in the smelting of other metals, notably aluminum and zinc. Also, the processes presently used to produce the single-crystal materials, such as epitaxial growth, are expensive and energy-intensive. The only reasonable way to generate large amounts of solar power with such cells is to use very small amounts of these scarce elements per watt of capacity. Two possibilities are thin-film cells and concentrator systems, such as the Boeing development mentioned above. It is impossible to tell if these cells will make a major contribution to the solar cell industry, but they will certainly be used in those situations where considerations of high efficiency or good thermal stability outweigh those of cost.

One use of photovoltaics in which cost is no object is military and space hardware. Gallium arsenide cells are now routinely used in space satellites and the Department of Defense is sponsoring research to investigate promising III–V compound solar cells. These grants, small by military standards, have become an important portion of the support that the U.S. government is giving to solar cell research.

## ELECTROPLATED CELLS

Many of the new semiconductor materials are extremely strong absorbers of light. They are called **direct band gap semiconductors** and the light that is absorbed to create the charge carrier is absorbed in the first micron or so of the material. This means that the entire solar cell can be extremely thin and can be made of polycrystalline material if the average grain size is larger than the micron or less that the charge carriers have to travel. If the grain boundaries can be **passivated** so that they won't act as carrier recombination centers, the crystal grains can be even smaller. (A **recombination center** is a point at which a hole–electron pair created by the absorbed light can recombine before the electron passes through the external circuit. It acts as an internal short circuit.) With thin layers of small grain size, electroplating is an excellent way to build a thin-film solar cell. Electroplated cells have been made with cuprous oxide, zinc phosphide, and cadmium telluride.

Cuprous oxide is called an "emerging material" by the planners in the Department of Energy even though it was the first known photovoltaic cell material and was studied in the early part of this century. It is a p-type semiconductor and simple Schottky barrier solar cells can be made by heating a sheet of copper in air for 15 minutes to 1000°C, followed by annealing at 500°C for another 15 minutes. This solar cell is extremely inefficient because the active junction is between the thin cuprous oxide layer formed and the underlying copper; the light must pass through the entire cuprous oxide layer to get to the junction. Several years ago, Dr. Dan Trivich and I developed a method of making front-wall Schottky barrier solar cells in which the active junction is between cuprous oxide and a thin semitransparent metal layer covering its surface. These cells display an efficiency of only 2%, but our theoretical studies indicate that an efficiency of over 13% may be possible, if the right combinations of materials and processing steps could be found. It is possible to electrodeposit cuprous oxide thin films in a process that would cost pennies per square foot; if cells of only 5% efficiency could be thus manufactured, costs of $0.25 or less per watt of capacity would be achievable. The supporting substrate would be the most expensive item in the finished product.

Similar costs per watt would be possible for electroplated cadmium telluride and copper indium diselenide solar cells. With electroplating, it is possible to have precise control of the materials used so that alloy depositions are possible. This control also means that electroplated solar cells can use heavy metals efficiently and safely with no toxic waste disposal problems in a well-run plant. Although several groups are hard at work on electroplated solar cells and laboratory results look very promising, we should not expect commercialization of the process for several years.

Many combinations are possible and a number are already under investigation on a small scale. The main barrier to developing inexpensive, efficient solar cells is a lack of knowledge of the semiconducting properties of these promising materials.

## ORGANIC SEMICONDUCTORS

Among the little-explored potential solar cell materials are organic semiconductors. These materials have been used in experimental photovoltaic cells in two ways—as photosensitizers and directly as the semiconductor element in a cell structure. It has been known for a long time that certain classes of dyes can, if deposited in a thin layer onto a base material, make that material photosensitive to the light the dye absorbs. This technique has been used for nearly a century now by the photographic industry to make orthochromatic and, later, panchromatic film. The silver halides used in photographic film (silver chloride, bromide, or iodide) are only sensitive to blue or ultraviolet light, but the dyes used to coat the tiny crystals of silver halides can absorb the rest of the visible spectrum and somehow transfer the energy of the absorbed light to the underlying crystal. It is now generally believed that this transfer involves the actual transfer of a charge carrier (either a hole or an electron) to the substrate material. The same technique has been commercially used in copying machines. Amal Ghosh at Exxon, and others, succeeded in making photoelectrochemical cells that can produce both hydrogen and electricity by dyeing titanium dioxide–coated electrodes with the organic dye phthalocyanine, thus rendering them sensitive to the visible sunlight that titanium dioxide (the main constituent of good quality white paint) normally reflects. The efficiency of such cells is still low, but they appear to be quite stable and would be very inexpensive to produce.

Organic semiconductors have also been used directly in Schottky barrier photovoltaic cells. About twenty years ago, while experimenting at Xerox Corporation with layers of phthalocyanine and other organic dyes in an attempt to determine the mechanism of energy transfer between the dyes and various substrates, I accidentally discovered that some of the devices exhibited a photovoltaic effect. We studied this effect for a while since it looked promising as a photodetector for Xerox's high-speed facsimile system then under development, but nothing was done at that time to exploit the system for solar energy conversion. Xerox of Canada has taken up the idea as a potential solar cell, but the efficiency achieved so far is still low because of the poor conductivity of the material. But this class of organic pigments, all similar to chlorophyll in chemical structure, could produce an extremely cheap solar cell that could be literally painted onto the substrate.

A new way to use dye-sensitized inorganic oxides has been discovered by Michael Gratzel of the Swiss Federal Institute of Technology. He produces a photoelectrochemical cell by coating titanium dioxide with special ruthenium containing organic dyes. These dyes bind extremely well to the colloidal titanium oxide, allowing him to prepare very thin, highly absorbing films onto the surface of tin oxide–coated glass. The layers are very stable and allow the conductively coated glass to be used as a window electrode in a liquid cell. A thin opaque platinum counter-electrode forms the back contact. The electrons excited in the dye molecules by the energy of the absorbed photons have only a very short path through the titanium dioxide to the degenerate n-type tin oxide which becomes the negative electrode of the cell in the light. Electrons then flow through the load and are reinjected into the ionic liquid electrolyte from the back contact. The active dye molecules take up the electrons to complete the circuit.

Gratzel and his associates recently devised a totally solid-state version of this photovoltaic process, and a newly formed company, Sustainable Technologies Australia (STA), is manufacturing pilot quantities of thin film cells on a glass substrate. The PV cells now produced are below 6% efficiency, but company scientists hope to achieve above 10% efficiency when in full production. These elegant, transparent, burgundy-red PV modules are well-suited for building-integrated PV walls and skylights.

Recently, Alvin Marks, the inventor of 3-D movies, has been working on a new concept for an organic solar cell. He has been taking the thin polarizing plastic films used in those cheap paper glasses we wore to watch the movies and changing the formulation of the linear polymers to make the films electrically conducting. The idea is that the films absorb virtually all of the light of a particular polarization and produce free electrons that will be captured at an electrode strip on the film. Two films with crossed polarization will absorb all the incoming light, leading to a high quantum efficiency.

Alvin Marks started Phototherm Inc., a company in western Massachusetts, to work on this idea. His basic idea is feasible from a scientific point of view. With the help of a couple of small grants and some research assistance from Argonne National Laboratories, he has produced the conductive polarizing polymer films and has produced a small voltage under illumination, but the cell resistance is still too high for a commercial device. The tricky steps seem to be those of incorporating the necessary diode junction into the polymer chains using charge transfer dyes and arranging a path for the electrons to get to the collecting electrodes without shorting out the microcells.

Another embodiment of his idea, using microelectronic techniques to fabricate the multitude of submicron-sized dipoles onto an inorganic substrate, is being tried at the nearby University of Lowell (Massachusetts). This system should produce a working device but will probably be too expensive for ordinary use.

Alvin Marks has been claiming very low cost and high efficiency for his Lumiloid polymer solar cell. While the plastic film cells will probably be inexpensive to produce, quantum efficiency (which can be very high for organic dye molecules) differs from device energy efficiency (which will probably remain low). However, an efficiency of only 5% would be commercially useful in a simple device.

There are a number of other dark horse solar cell possibilities that have been given only cursory research efforts. Any one of these, if proven to be commercially usable, could suddenly change photovoltaics from a rarity used only at remote sites and for special situations to the system that replaces all conventional utilities. Chlorophyll membrane systems, photosensitized zinc oxide, and little understood compound semiconductor systems fall in this category.

## CLEVER OPTICAL SYSTEMS

In addition to exploring new solar cell materials, researchers are also looking hard at ways to make existing solar cells utilize sunlight more efficiently. Most of these systems employ clever optics to direct or focus the light before it is absorbed by the cell.

The simplest systems are the hybrid optical concentrator systems discussed in Chapter 3. New inexpensive Fresnel lens optical systems, nonfocusing concentrators that allow large tracking errors, and automated assembly systems promise to lower the final installed cost of concentrator systems and make them competitive with the cheapest flat plate photovoltaic collectors. Most of these designs incorporate methods to extract the heat generated in the solar cell and put it to use, making the system even more economically attractive. SunWatt has introduced an inexpensive hybrid concentrator based on an aluminum fin with a copper tube, similar to the design of a solar water heater.

## CASCADE CELLS

The cascade cell, first proposed in 1953 by Dan Trivich, is simply a stack of different solar cells. Each one absorbs part of the solar spectrum and passes the rest of the light on through to the other cells in the stack. The cell made with the semiconductor that has the largest band gap has the shortest cutoff wavelength and transmits most of the light redder than this wavelength. This light can be utilized by a semiconductor with a narrower band gap placed behind the first cell. It is possible to stack even a third cell or more and efficiently use the entire spectrum of sunlight including the infrared. A single, narrow band gap cell would also absorb the entire solar spectrum, but would have a low output voltage, essentially wasting a good deal of the energy of the short-wavelength light. The cascade cell will have an output voltage that is the sum of the voltage of the individual cells. The output current of the device is less than that of the simple, narrow band gap cell, but the overall efficiency ends up much higher. Experimental two-cell devices have been constructed using gallium arsenide and aluminum gallium arsenide and various combinations with silicon, gallium arsenide, and germanium. In some of these systems all the various cell layers are constructed directly on top of each other; they utilize a **tunnel junction** or some other sort of shorting junction to make an electrical connection between the cells in the device. Actual efficiencies of over 33% have been measured and theoretical efficiencies of 60% or more have been predicted for these multijunction devices.

The Department of Defense is interested in these **tandem cells** because of the great amount of power delivered by a small package. A photovoltaics-powered unmanned airplane with two-layer tandem cells covering the wings is now a possibility. The cells would produce enough extra power, stored in the batteries, to keep the plane flying all night. The drone could circle for months at high altitude, taking surveillance pictures.

Most experimenters envision using the cascade cell at the focus of an optical concentrator system since the cost of the complex cell is expected to be high and the efficiency of the system actually increases with greater light intensity.

Another version of this is shown in Figure 6.4. Here the incoming light is split into various colors with selective filters and directed to several different cells, each is designed to use a particular band of wavelengths more efficiently. Varian Associates has also been working on a version that uses high concentration ratios and very efficient dielectric-coated filters as the beam splitter. A potentially cheaper

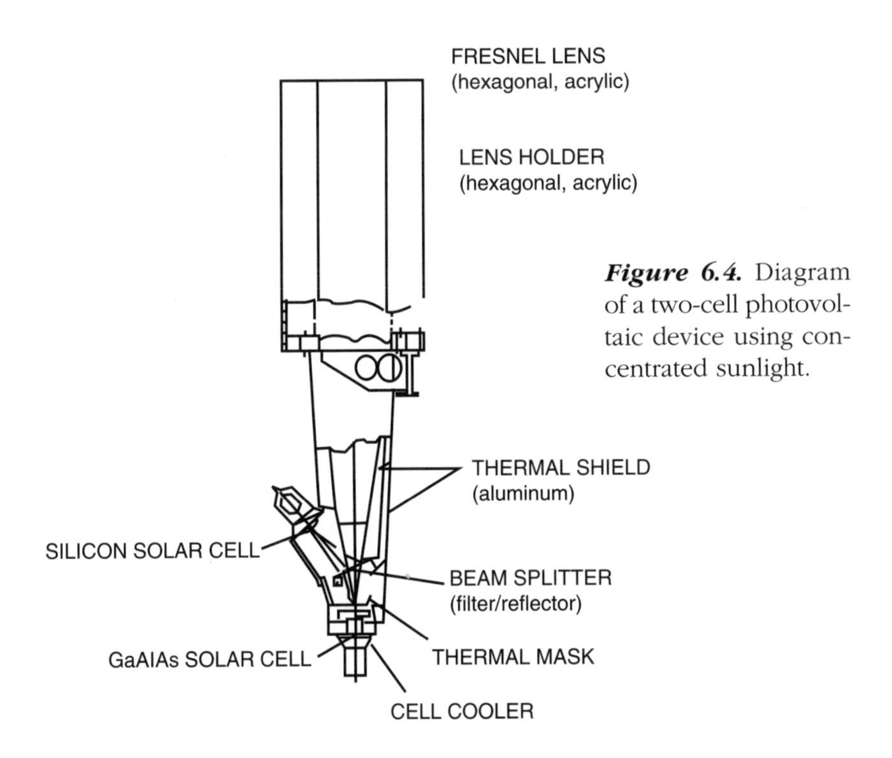


***Figure 6.4.*** Diagram of a two-cell photovoltaic device using concentrated sunlight.

version is being developed by W. H. Bloss at the University of Stuttgart; it uses holograms developed in gelatin as the dispersive concentrating (DISCO) system. The DISCO system focuses the light and splits it into several bands of different wavelengths directed at the two or three different solar cells used.

## LUMINESCENT CONCENTRATORS

One of the most exciting new concentrator systems is the luminescent solar concentrator which absorbs solar radiation in a flat plate by means of colored dyes. These dyes fluoresce in all directions at a longer wavelength than the incoming radiation. Most of this reradiated light is confined to the collector by total internal reflectance and transmitted to the edge of the sheet where small photovoltaic cells convert it into electricity. The basic concept has been known for years and, in fact, has been used in drafting triangles for illumination. The application to solar energy collection was first proposed in 1976 by Lambe and Weber of Ford Motor Co. Research Labs. Since then a number of groups have worked on the idea, using organic dyes in plastics or inorganic fluorescent compounds in glass. A rather complete theoretical treatment of the efficiencies and concentration ratios to be expected is given by Goetzberger and Greubel (see Recommended Readings).

Wood and Long experimented with liquid-filled flat plastic cases as luminescent concentrators in which a mixture of dissolved dyes absorbs most of the visible solar spectrum. A cascading of energy, which they envision as a series of fluorescences and reabsorption by the different species of dye molecules, produces a final fluorescent light which is transmitted at a long wavelength to the photocells at the edges of the case. In their experiments they found a rapid heat build-up in the liquid because all the incoming light energy that is not finally converted to electricity or that escapes from the case is trapped as heat. They also found that the dyes bleached rather quickly under their experimental conditions.

At Skyheat Associates, I have been working on a version of a luminescent concentrator that utilizes both the heat and the electricity developed from the sunlight. Dyes dissolved in a liquid absorb the solar radiation and transfer the light to solar cells fastened to the

***Figure 6.5.*** A luminescent concentrator. This flat plastic tank contains a colored fluorescent liquid. The liquid concentrates light onto small solar cells glued to the edges of the tank. (Skyheat Workshop photo)

end of the flat tank. The liquid is heated and can be pumped through a heat exchanger to extract the heat generated. Cooler operating temperatures and a careful choice of dyes seem to greatly reduce the dye stability problem, but the system is still in early development and working models are too inefficient to be of practical interest.

## THE SOLAR THERMOPHOTOVOLTAIC SYSTEM

This complicated-sounding name describes a system that combines some aspects of solar thermal concentrator devices with a method of converting the heat generated by focused sunlight directly into electricity. If a large parabolic mirror focuses sunlight through a small opening into a well-insulated black cavity, the interior of the cavity can become extremely hot (1500°C or more). This temperature causes the inside walls to radiate in the infrared, with the radiation bouncing back and forth in the cavity. If a germanium photovoltaic cell is placed in one wall of the cavity, the infrared absorbed can be converted into electricity because of germanium's narrow band gap.

Loferski and others at Brown University developed a special, high-efficiency germanium solar cell especially for this application and calculate that overall conversion efficiencies as high as 20% are possible with such a system. The germanium cell would actually stay cool, even in such a hot environment, since the incoming infrared energy is converted into electricity. However, if the cell were disconnected from the electric load, it would quickly heat up and destroy itself. The solar thermophotovoltaic system has the disadvantage (or advantage, depending on your viewpoint) of only being practical in a large centralized system like a **power tower** where economy-of-scale justifies the expense of the complex tracking system for a whole field of steerable mirrors. Recently, Applied Solar Energy started work under government contract to develop these high-efficiency thermophotovoltaic cells.

## RECOMMENDED READINGS

Bloss W., J. Loferski, et al., eds. *Progress in Photovoltaics.* 1:1 (January 1993) (Sussex, England: Wiley & Sons).

Goetzberger, A., and W. Greubel. "Solar Energy Conversion with Fluorescent Collectors." *Applied Physics* 14:123–39 (1977).

Hovel, Harold. "Photovoltaic Materials and Devices for Terrestrial Solar Energy Applications." *Solar Energy Materials* 2:277–312 (1980).

*Proceedings of the IEEE Photovoltaic Specialists Conferences* (New York: IEEE, January 1980 though May 1993).

Zweibel, Kenneth. *Harnessing Solar Power: The Photovoltaic Challenge* (New York: Plenum Press, 1990).

# Chapter 7
# The Future of Photovoltaics

The future of photovoltaics looks bright. A viable industry now exists and myriad uses for photovoltaic devices have been discovered and developed. Product lines have been standardized and in 1997 PV companies shipped 120 megawatts of photovoltaic modules worldwide. A secondary market in **balance of systems (BOS)** equipment has emerged to furnish the non–solar cell components needed for a complete installation. A number of firms specialize in designing and installing finished systems for remote sites all over the world. Corporate giants such as General Electric and Westinghouse now have photovoltaic divisions working on these installations; the industry is fast maturing with well-trained electrical engineers designing reliable systems that compete in remote places with diesel generators. (In some cases, hybrid PV/diesel systems are chosen as the low-cost answer for remote power supplies.)

As the prices of cells and modules drop, the applications for which solar cells are the best choice will greatly increase and the market will grow exponentially. The prices will decrease over the long run, not just because the expanded market allows for more efficient production but because research and development will produce less expensive photovoltaic devices. The speed at which innovation and development of production capacity will occur is sensitive to decisions made by government agencies, and influenced by the acquisition of small innovative companies by large oil companies and other multinationals. Obstacles will arise as economic pressures keep the price of solar cell modules from dropping as quickly as desired and the lack of capital keeps small companies from implementing their automation and expansion plans. But these plans will become reality, new types of solar cells will become available, and we will be using them increasingly in the near future.

## U.S. GOVERNMENT AND PHOTOVOLTAICS

Since the space program began in the late 1950s, the U.S. government has been the world's major purchaser of solar cells. Most photovoltaic research and development has been financed by the government; in fact, the present-day photovoltaic cell would probably not exist except for this commitment. The expenditure of tax money on solar cell research has not been a matter of altruism, but rather the purchasing of a technology that was, and will continue to be, vitally important to our space program, and one that is fast becoming important in meeting our modern military needs. Our commercial satellite communications network would not exist without this commitment to photovoltaic research. In the future, the government will use ever larger quantities of solar cells for both space and terrestrial applications, as photovoltaics proves to be the most economical, reliable, and, quite often, secure way to generate electrical energy in many applications.

Since the formation of the U.S. photovoltaic program in 1974, the government has clearly defined the goals for the emerging photovoltaic industry. The Department of Energy established guidelines for the development of terrestrial solar cells that, if met, would result in solar cells replacing 1 quad of fossil or nuclear fuel by the year 2000 (1 quad of $10^{15}$ Btu or $2 \times 10^{11}$ kWh). The DOE has hopes that lowering the cost of solar cells will stimulate the market.

Figure 7.1 shows government projections for solar cell costs. The $2.00 per peak watt predicted for 1983 was calculated as the

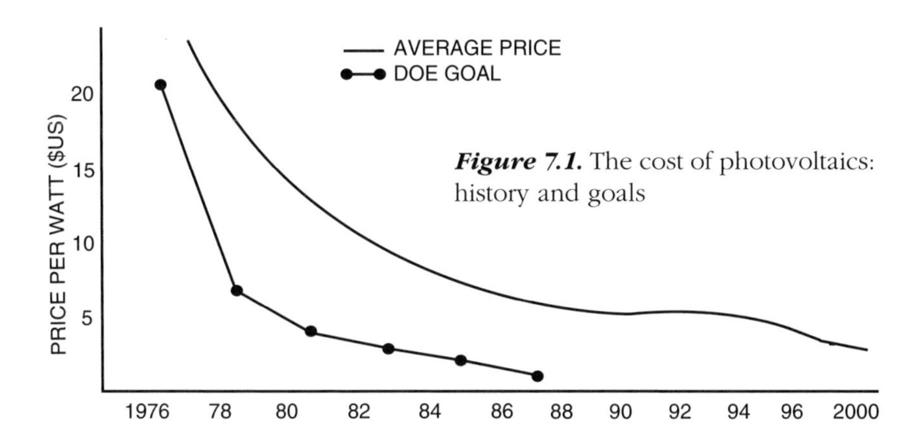


***Figure 7.1.*** The cost of photovoltaics: history and goals

price necessary to attract users of electricity in remote areas. While that goal was not met, these users proved interested in solar cells, even at present prices.

A great deal has been written about U.S. government involvement in photovoltaics and about the relationship between price and market penetration. (See the Recommended Readings at the end of this chapter.) During the Reagan and Bush administrations, the government's interest in solar energy was curtailed, but not totally eliminated. In fact, in the latter years of the Bush administration, photovoltaics once more became important as a reliable and environmentally friendly source of electricity. For instance, during this time, the U.S. Coast Guard undertook a program of utilizing PV modules to power all of the lighted marker buoys and remote lighthouses in this country. Most other countries are following suit. The recent realization of the importance of sustainable energy sources has reinforced the government's interest in meeting the goals shown in Figure 7.1.

The National Energy Research Labs (NERL) [formerly the Solar Energy Research Institute (SERI)] in Golden, Colorado, had a very active photovoltaics research program. Although constrained by budget restrictions, NERL has an active interest in the compound semiconductor solar cell as well as those systems which are too new or poorly developed to interest profit-oriented business, as well as implementing basic research into the physics and chemistry of photovoltaics. Even though the regional solar centers formed with SERI have been phased out, the NERL serves an important function as an information source to the financial and business communities as well as to fellow scientists.

The Department of Energy has started the Photovoltaics Manufacturing Technology (PVMaT) project as a government/industry partnership to research and develop photovoltaics manufacturing techniques. This program is examining ways to reduce manufacturing costs, to make PV modules more reliable, and even to recycle the materials from obsolete or deteriorating modules. The final goal is to increase the U.S. production capacity and lower the cost per watt.

## THE MULTINATIONALS AND PHOTOVOLTAICS

Many companies that manufacture solar cells are now owned by oil companies. There has been considerable debate about whether this is good or bad for the industry and for the ultimate consumer. Before we condemn the trend outright as another example of big business monopolizing a good thing, we should examine it in detail.

The multinational oil companies see themselves as energy companies and, after gaining control of most coal and uranium mines, the natural step would be to move into solar energy. This move has not been without difficulties; easily attainable, centralized resources do not exist in the solar field. The inherent decentralization of solar and most renewable energies does not fit into oil companies' existent distribution and marketing structure. Since solar energy must compete with the well-established and heavily subsidized conventional energy sources, the profit margins are very slim.

Why, in the face of these difficulties, would the oil companies want to control this industry? As our strong dependence on petroleum is forced to an end, the oil companies must cover all the options if they are to continue as the world's energy suppliers. Given the size of the solar cell industry and the capital resources available to the oil companies, it is easy for them to buy fledgling solar cell firms. The oil companies know that the rising cost of conventional energy, the present delivery prices of photovoltaic cells, and the promise of a breakthrough in solar cell technology will lead to an enormous growth potential for photovoltaics—and they want to participate in that growth.

The real question is, is this good or bad? Is it helpful or harmful or even possible for a few large companies to control a world energy source? And will it be good for the solar cell companies to be run as small parts of giant corporations?

The theory that oil companies buy solar cell companies to put them out of business and make photovoltaic cells unavailable is unrealistic. This would be impossible to do. Solar cells have already established themselves as indispensable in certain applications—space satellites and lighted marine buoys, for example—ensuring that their manufacture must continue. In addition, the technology for making photovoltaic cells is so well-established and the world market so large that suppliers to the terrestrial market could enter the field faster than oil companies can buy them out. There is no evidence that the solar cell companies already owned by oil companies are not attempting to produce better and cheaper solar modules and to expand their markets. Occasionally, an entire company's product line does disappear, as did Exxon's Solar Power products. But rather than an attempt to make solar cells unavailable, this was a demonstration of the power of unlimited capital resources. A giant corporation can simply stop selling a product for years, if necessary, and absorb the operational expenses of an entire division while developing a new product line and production facilities.

Large amounts of research and development money for new photovoltaic systems and the capital resources to quickly scale up production with automated assembly lines could be a boon to the solar cell industry. The government need to control federal funding for solar cell R&D makes the cash available from large corporations even more desirable and, some claim, necessary. So it is possible that, given the present situation, oil company ownership of the solar cell industry could do a great deal of good. They have the resources to expand the industry much more rapidly than small companies that must bootstrap themselves on slim profits. Venture capital for a new technology like photovoltaics is difficult to obtain from conventional sources at present. The oil companies do have the capital and they do appreciate the importance of photovoltaics.

Since the first edition of this book was published, Solarex has been completely bought out by Amoco (Standard Oil of Indiana). Amoco had a minority position in Solarex for years, but in 1983 circumstances led Solarex to have cash flow problems and the oil company took the opportunity to acquire the remainder of Solarex stock at bargain prices. Two events led to the takeover. First, when RCA decided to give up its amorphous silicon research and development, Solarex saw the chance to acquire the leader in the field. They correctly perceived RCA as the largest potential threat to the Semix polycrystalline silicon process. Second, Solarex had just borrowed heavily to construct the Solar Breeder facility in Maryland. This innovative solar cell factory, powered by a roof covered with solar cells, gave Solarex the capability to make photovoltaic cells from scratch (that is, starting with sand) without any outside power source. Unfortunately, at the time the Solar Breeder was finished and dedicated, sales of solar modules were down and the cash flow was not sufficient to pay off the loans. Amoco supplied the cash. Since the merger of British Petroleum and Amoco, Solarex has now been merged into BP Solar, which is now the largest photovoltaics manufacturer in the world.

The oil companies themselves provide one large market for solar cells. They know what it costs to furnish electric power to remote sites and now use photovoltaic arrays on oil drilling rigs (see Figure 7.2) and for corrosion protection devices on pipelines. A parent oil company could probably use the entire output of its solar cell division, thus giving the division incentive to expand and produce greater quantities of inexpensive solar modules. It is hoped that at some point the solar cells would be sold to the general public, passing along, of course, the cost savings.

Arco Solar, once the largest manufacturer of photovoltaic cells and modules, sold its entire operation to Siemens of Germany. Siemens has invested considerable money in upgrading and expanding the former Arco Solar manufacturing facility in Chatsworth, California, and is exporting more than 20 megawatts of single-crystal silicon cells to module assembly plants in Germany and elsewhere. Siemens is also continuing the former Arco Solar research effort in thin-film solar cells.

It is impossible to predict long-term results of the oil company–solar cell connection. Given the resources, solar cell companies could grow and become a very important part of the energy industry, furnishing virtually all our electric power needs at a reasonable price. The nagging concern is the advisability of allowing this important energy source to be owned by the same organizations that control our other energy sources.

***Figure 7.2.*** Photovoltaic modules are being used by oil companies on offshore drilling rigs to operate the marker lights and communication equipment. (Solenergy Corporation, Woburn, MA)

## PHOTOVOLTAIC POWER FARMS

Ever since the nineteenth century when the amorphous selenium cell was developed, people have dreamed of photovoltaics' potential as a major electric power source. Now with the existence of **solar power farms**—large racks of solar cells that generate electric power in remote areas and then transmit it to the power companies—the dream is being realized.

The first large-scale photovoltaic power systems were United States government demonstration projects: Indian villages in the southwest, an irrigation system in Nebraska, even a radio station in Ohio were powered by modules that represented a major portion of all the solar cells produced in the mid-seventies. These demonstration projects gave the industry a needed boost and produced a database of operating experience and module failure modes that has proved invaluable in the search for more reliable designs.

Large solar cell power modules were next applied in areas where conventional power was unavailable and the cost of extending the utility grid prohibitive. Villages in Saudi Arabia and Africa were fitted with these experimental systems, each one a custom design. The Saudi system is particularly interesting in that it incorporates long racks of Fresnel lens concentrators to focus the sun onto specially designed photovoltaic cells. The experience in designing the controlling and tracking mechanisms has lead to new levels of sophistication in electronic circuits and drive mechanisms. The array has a peak output of 350 kW.

All these projects were subsidized at first by a government. It was believed that it would be years before photovoltaic cells became cheap enough to justify their large-scale use in competition with more power sources. But the passage of the Public Utilities Regulatory Powers Act (PURPA) by the U.S. Congress changed the economics. It is now possible for a small producer (less than 80 megawatts) to install a generating system and sell the power to the utility at a favorable price and without the enormous amount of red tape usually required of a new electric power producer. This newly updated act plus the renewable energy and investment tax credits (for business) have made renewable energy attractive as an investment. In California, wind power farms have been running for years in several famously windy passes, and small hydroelectric plants are being built along almost every suitable stream. Investors have included photovoltaics in their plans and solar power farms are now a reality.

In December 1982, the first photovoltaic power farm went into operation at Carrizo Plains in California. Built by Arco Solar, this power farm grew to several megawatts in size and incorporated an unusual set of tracking units that allow a set of flat plate modules to follow the sun for about eight hours every day. Arco Solar had included concentrating mirrors in a large section of the farm. These flat mirrors doubled the output of the PV modules, but over the six or more year of operation, the extra heat started decomposing the ethylene vinyl acetate encapsulant films bonding the solar cells to the front cover glass. Arco did not include the Carrizo power farm in the sale to Siemens and it has since been dismantled; this has lead to a great number of Carrizo-Gold, -Bronze, and -Copper PV modules on the surplus market.

Arco Solar also won the bid to furnish the first sections of a very large photovoltaic power plant being built by the Sacramento Municipal Utility District (SMUD). The twelve-year project is expected to furnish at least 100 MW when finished and should produce power for less per kilowatt hour than the nearby Rancho Seco nuclear power plant. The power farm includes samples of amorphous silicon PV modules as well as the more conventional crystalline silicon modules. The Sacramento utility project is being financed partly through the Department of Energy, money they intend to repay when the average price per watt of modules in future purchases becomes less than $3.20.

Numerous solar power farms were rushed to completion before the solar tax credit for individuals expired in 1984. It had been hoped that by the time this happened, the price of solar cell modules would have dropped to the point where investment in large central systems made good sense without the artificial support offered by the tax

***Figure 7.3.*** PVUSA, Davis, California.
(Siemens Solar Industries, Camarillo, CA)

credits. Although the price did not drop to the extent hoped, solar cells have become cost-effective in many applications.

An important new PV power farm is Photovoltaics for Utility Scale Applications (PVUSA) in Davis, California. This is a national cooperative research project established in 1992 by DOE, the Electric Power Research Institute (EPRI), and a consortium of electric utilities. The power farm includes a set of 20-kW arrays from many different manufacturers representing what they consider their "emerging module technologies." These include advanced crystalline silicon modules, such as some using microgridded single-crystal cells from Siemens Solar, bifacial polycrystalline cells from Solarex, the thin polycrystalline cells on substrates by AstroPower, and ENTECH's two-axis linear concentrator system. Also included are several different amorphous silicon arrays and large 200- to 500-kW arrays that use more conventional modules, some with passive tracking systems. All of these arrays are instrumented for data collection and connected to the Pacific Gas and Electric (PG&E) utility grid.

EPRI has issued a series of reports over the years evaluating photovoltaics as a potential electrical generator. Though they only consider systems from the utility's point of view—downplaying distributed solar cell arrays on individual homes, even those that would be grid-connected, their study of the economics of the centralized solar power farm is informative and carefully done. They conclude that solar cells must be both cheap and efficient before a utility could justify considering them as an alternative to coal or oil. Nuclear power plants were not mentioned in this report; their economics compare quite unfavorably, of course, to the alternatives. Because of the very high balance of system costs assumed for the power farms, the conclusion is that only advanced cell types or concentrator systems have any chance of economic acceptance. In spite of the pessimistic view of the utility industry, plans are moving ahead for more and larger photovoltaic power farms. Most of these will use flat plate modules, but some will incorporate concentrators and cogeneration.

In the future, the photovoltaic roof will become more common on homes that are part of a distributed solar power farm. This will save a good portion of the balance of system costs related to both the acquisition of land and the building of support racks to hold the thousands of modules required. These distributed power farms can be organized and financed in three ways:

1. The individual homeowner may decide to invest in an array, expecting that the savings in utility bills and the small return each month from the utility company to eventually pay back the cost of the system. This is the scenario examined by EPRI. If the slow acceptance of solar water heaters with their potentially much faster payback time is any indication, this is the least likely to succeed.

2. The utility itself may invest in a large number of small solar cell arrays mounted on its customers' homes. By simply renting roof space through some equitable arrangement, the utility would save a substantial portion of the balance of systems costs. The Sacramento Municipal Utility District and Idaho Power have already initiated such programs and the idea is being seriously considered in many other places. This option should be particularly attractive to small rural electric coops that are member-

***Figure 7.4.*** Delmarva Power & Light installation, Newark, Delaware. (AstroPower, Newark, DE)

owned and must currently purchase all their power from large utilities. That the members produce their own power is much more logical than their investing in a nuclear power plant that might never get built.

3. Independent third-party investors could put up the capital for systems installed on an entire subdivision under construction. The homeowner is compensated for the use of the roof and the power is sold directly to the utility. Such a financial package would work particularly well with a solar module such as the SunWatt panel with water-cooled cells. These hybrid modules make hot water and electricity at the same time, so the homeowner would rent the array for the hot water output while the third-party investor would receive all the revenue from the sale of the electricity.

Any of these distributed systems would require multiple inverters and controls to ensure safe and reliable operation, but these inverters are now readily available off-the-shelf items. In fact, in the near future the utility tie operation will be offered as standard on any large inverter. The main impediments will be political and social: People will have to get together and boldly work out the plans for our future use of renewable energy to furnish the electric power so necessary to the good life we desire.

## FOREIGN SOLAR CELL ACTIVITY

The photovoltaic industry is international in every sense of the word. Not only are solar modules used in every country, but major contributions to the development of photovoltaics are made by private companies, universities, and government laboratories worldwide. To list all the programs would be impossible. Some important new developments, however, should be mentioned.

### Japan

Next to the United States, Japan has the greatest commitment to research and development of solar cells. The goal of the Japanese

***Figure 7.5.*** 350-kW photovoltaic power farm is furnishing power for remote villages in Saudia Arabia. Martin-Marietta designed and built the tracking concentrator arrays. (David Buzby photo)

government's "Sunshine Project" is to reduce Japan's dependence on imported fuels by the extensive use of solar energy. The close cooperation between government, academia, and private industry has already led to the rapid development of new amorphous silicon solar cells. The *a*–silicon carbide–silicon cell is an outstanding example of this technology. At the present time, all commercially available Japanese PV modules are made from crystalline silicon. All available Japanese amorphous solar cells are the small ones used for pocket calculators and digital watches, but Japanese companies should be selling amorphous silicon solar cell modules and arrays at extremely competitive prices very soon. Kyocera, the ceramics company, has been using new crystal-growing techniques to produce the polycrystalline silicon solar cell modules they sell in the United States. Sumitomo Corporation and Sanyo jointly acquired Solec International, the third largest solar cell manufacturing company in the U.S and one of the oldest companies in the PV industry.

**Western Europe**

The photovoltaic cell was invented by Frenchman Edmond Becquerel in 1839 and French scientists have never lost their interest in solar energy. In past years, the French government had been less than excited about the large-scale development of photovoltaic cells (possibly because of their heavy commitment to nuclear power). Since the advent of the concerns about global warming and nuclear waste, however, the emphasis is shifting toward renewable energies and new R&D efforts in photovoltaics. Individual French scientists and companies working on low-cost systems have been continuously in the forefront of solar cell research. French companies seem to get into the photovoltaics business by buying small American manufacturers rather than by setting up their own facilities. This trend may be changing since two French oil companies have dropped their American subsidiaries and are now starting production on their own PV assembly lines.

The German government has begun to develop and encourage the use of photovoltaics and a number of German university scientists, including Bloss in Stuttgart and Goetzberger at the Fraunhofer Institute, are exploring new photovoltaic systems. Until recently, Germany's commercial development of photovoltaics had been largely in furnishing raw silicon materials for solar cell manufacture. The Siemens process of purifying silicon for the semiconductor industry has become the standard. Telefunken pioneered the method of casting cubes of polycrystalline, solar-grade silicon. These companies are expanding their production facilities. Siemens greatly expanded its American facilities (the former Arco Solar) and now owns the world's largest PV manufacturing facility. They are developing new lower-cost processes and definitely anticipate capturing a large share of the growing photovoltaics market. They envision that their greater market will be in the underdeveloped countries.

Since the Kyoto Agreement, Europeans have become very serious about manufacturing and using photovoltaic devices, particularly in building-integrated systems. BP Amoco has just finished constructing a large PV cell assembly plant in Spain, and Shell has built a large photovoltaic plant in Germany. This plant, with its spectacular curved photovoltaic roof, has room for up to four 20-watt production lines, one of which is already in operation. The Italian manufacturers are developing new PV-powered devices, as well as producing solar cells and modules.

**Underdeveloped Countries**

Because they are currently so useful for powering small electric systems in remote locations, solar cells should make a great impact on countries that do not yet have central utility grids. These countries are taking a great interest in the progress of photovoltaics. Moreover, the World Bank and other international organizations are concerned with the tendency of underdeveloped countries to spend an ever greater part of their gross national product—and to go deeper into debt—to finance their energy needs.

Underdeveloped countries can began using solar cells in two ways. The easiest is to simply import and use solar modules made elsewhere. The market thus created is becoming a major one for American photovoltaic manufacturers. Solar irrigation pumps have been developed, notably by the French for use in North Africa, and solar-powered television sets are enabling remote villages to receive educational programs from satellite stations. For dispersed uses of

***Figure 7.6.*** PV-powered lighting installation atop a hospital in Fiji.
(Solarex Corporation, Frederick, MD)

electric power, solar cells (and wind generators in the appropriate climates) make much more sense, both from an economic and a cultural point of view, than dependence on large centralized nuclear- or fossil fuel-powered electric plants. The cost of running power lines through the wilderness and the problems of maintaining such systems would put an intolerable burden on the resources of most developing countries. It is likely that solar cells will be used widely in the Third World long before they are cheap enough to be practical for the American homeowner.

The second way a country can benefit from photovoltaic cells is to participate in their development and manufacture, either for internal consumption or for export or both. A number of countries—including Mexico, China, and Yugoslavia—have initiated photovoltaic programs to meet their internal needs and have established plants to make silicon solar cells and modules.

It is important in a rapidly changing field like photovoltaics that a country not get locked into an obsolete technology and that development plans remain flexible enough to take advantage of breakthroughs in low-cost cells. Scientists and engineers in a country with a healthy research program, even if the program is small, can actually be in the forefront of such a new field and can make significant contributions.

Many of the new types of photovoltaic cells now being developed, such as amorphous semiconductor cells and compound heterojunction devices, use simpler manufacturing techniques than those necessary for single-crystal cells. So it should be possible to set up relatively small manufacturing plants in developing countries to fill the energy needs of an area without disrupting the country's economy or requiring a cadre of highly skilled foreign technicians. Companies already engaged in the assembly of automobiles or the manufacture of other consumer goods will soon be able to tool up for the manufacture and assembly of solar devices, such as hybrid photovoltaics/solar heating modules. Unlike so many fruits of modern civilization, which have been paid for by social disruption and the economic dependence of developing countries, solar energy promises to be a means through which such nations can truly develop independently.

Thousands of village-scale photovoltaic arrays are already in place in developing countries. Using revolving loan programs, villagers are able to purchase one or two modules to run a radio and lights; they pay for the installation with the money saved by not buying kerosene. There is a module assembly plant in Morocco that furnishes panels to all of North Africa. Grupo Fenix in Nicaragua currently manufactures complete PV systems from scrapped solar cells made in the U.S. Many Peace Corps volunteers purchase small SunWatt units to recharge batteries and run laptop computers in remote worksites. There are more than 100 megawatts of PV power in use in developing countries. This is probably the fastest-growing market for photovoltaics.

## THE SOLAR ELECTRIC HOME

Thousands of homes are already powered by photovoltaic arrays. Most of these are small systems installed on remote dwellings by homeowners who wish to power a few lights, a radio or television, but who are far from power lines. Some have connected their systems to the utility grid and engage in buy/sell arrangements with a power company. But increasingly we are seeing fairly elaborate systems installed by people who want a fully equipped household but choose not to deal with a power company.

With recent advances in inverter design and the ready availability of efficient low-wattage lighting and appliances, a subindustry of photovoltaic installers and balance-of-systems specialists has emerged. These experts can furnish reliable, electrical household systems that are in every way the equivalent of a conventional grid-connected system. In the event of storm-caused power outages, especially in rural areas, the reliability of a stand-alone photovoltaics-powered home may actually be greater than that of one connected to the grid.

The economics of stand-alone home power systems is becoming increasingly favorable. When you compare the system cost to the cost of conventional wiring and the power company's charges to run utility lines to the homesite, the price per kilowatt hour is about the same. If the site is some distance from the utility source—a quarter mile, for example—a stand-alone system will actually cost less than the $10,000 or so the power company will charge to extend lines.

Although sizing, designing and installing a PV system is more elaborate than wiring a conventional home, it is not too complicated. (*The New Solar Electric Home* gives detailed instructions.) In most

***Figure 7.7.*** The occupants of this completely energy-independent home in Southern Indiana use two SunWatt PV/hot water hybrid modules to furnish all of their electrical and hot water needs. (Skyheat photo)

***Figure 7.8.*** The PV-powered lawn mower predicted by *Business Week* in the 1950s is now a reality. (Skyheat photo)

parts of the country, when weather data are compared to electrical needs, one finds a bad short-fall in the winter. A larger photovoltaic array would be required to handle these few bad months. In many places, combining solar cells with a wind generator can produce a reasonably constant supply of power *and* allow a reduction in the size of the battery bank needed to store the power. The SunWatt computer programs for system sizing include an excellent method for determining the proper combination of sun- and wind-powered systems.

The person contemplating a self-sufficient home must reduce power consumption requirements to a fraction of that used by the typical American homeowner. This can now be done, however, with little change in lifestyle. Very efficient alternatives exist for most home appliances and quite often the quality of life will be enhanced by the changes. One advantage of a passive solar home, for example, is the delightful living space created. Add a set of hybrid photovoltaic/hot water modules and all your hot water needs, including luxuries, can be accommodated. This third edition of *Practical Photovoltaics* was completely written in an off-the-grid home and typed into a solar-powered computer over the course of a winter in Maine.

In the future, more and more people will choose the path of energy independence—and not simply to save money. Consumer resentment at being forced to pay for the mistakes of utility executives will drive more people to disconnect from the utility grid. Financing canceled nuclear power plants, and the growing awareness of the dangers of considering them in the first place, will lead to the realization that viable alternatives exist in the form of reliable renewable energy systems.

## SOLAR CELLS IN SPACE

From the beginning of the space program, solar cells have played an important role in powering satellites. The original Vanguard satellite's one-watt solar array performed flawlessly. Since then the space program and photovoltaic cells have become sophisticated to the point where multikilowatt arrays are routinely used on communication satellites. The high-efficiency, blue-sensitive silicon cell that has become the standard design for commercial terrestrial cells was orig-

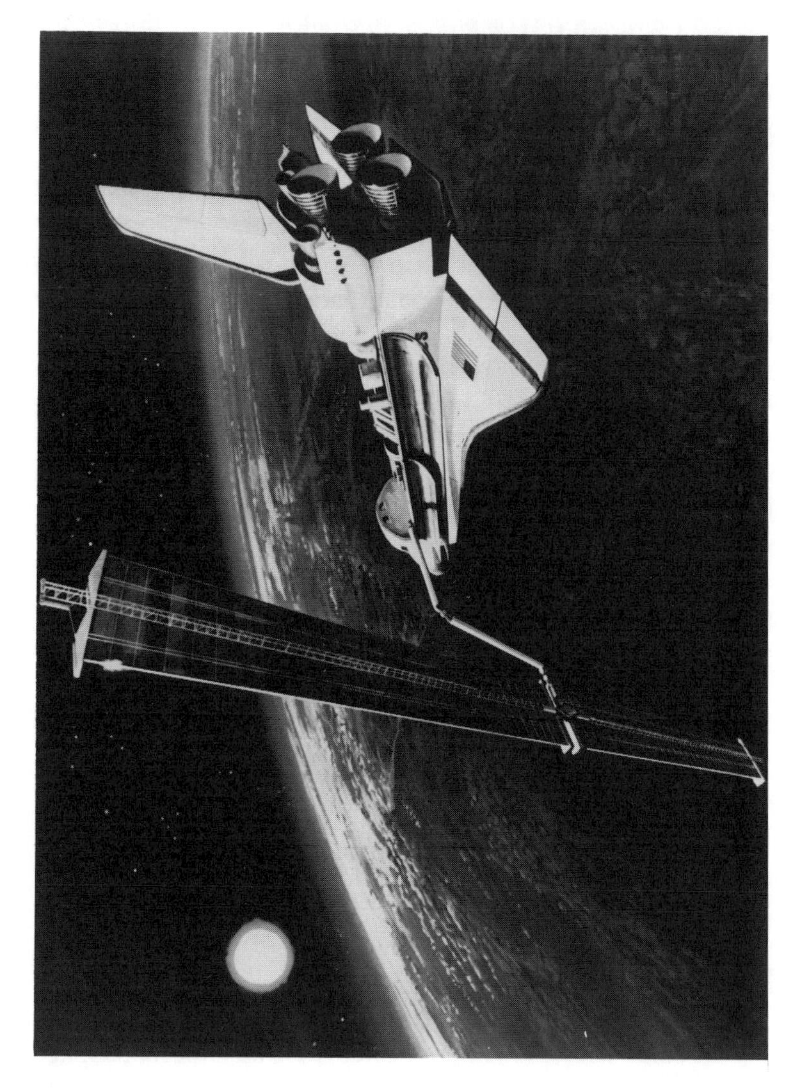

***Figure* 7.9.** The space shuttle carries a roll-up module to furnish electricity for extended missions.

inally developed by Joseph Lindmayer for space applications. The importance of quality control in manufacture and the consequent reputation for reliability that photovoltaic modules earned was part of the space program philosophy. Reliability will remain an important factor in the ready acceptance of solar cells for remote terrestrial applications, once the price becomes low enough to make them competitive with other small power sources.

With the advent of the space shuttle, the number of solar cells used in space is growing enormously. Researchers are working on new cell designs that will be even more reliable, more efficient, easier to assemble, and less expensive than those space cells now used. The space shuttle carries a large array that can be unrolled in space to extend the mission time long past that possible with the hydrogen fuel cells alone. The joint U.S.–Russian–European space station has the largest PV array ever launched into space (larger than the wings on a Boeing 747). It was designed to be left in space as a power station for the space shuttle.

## The Solar Power Satellite

One idea proposed for large-scale solar generation of electric power is the solar power satellite. First proposed by Peter Glaser, the solar power satellite is envisioned as a multimegawatt photovoltaic array parked in a synchronous orbit 22,000 miles out in space, beaming photogenerated power back to earth in the form of microwaves. This energy would be picked up by antenna farms, miles in diameter, either placed in remote land areas or floating offshore. A network of these satellites could furnish all the electricity our planet needs more cheaply and with much less environmental disruption than the equivalent number of nuclear or coal-fired plants. For proponents of the status quo, the solar power satellite system has the additional advantage of being a centralized source of electricity; the energy could be fed into the existing electric grid which could be owned and controlled by the present utility companies.

There are a number of safety and environmental questions concerning the effects of enormous amounts of microwave energy focused onto selected spots on earth—although the plan is to spread the beam over such a large area that its intensity is diluted to the point that no known harm would be done. Carefully designed feedback controls would ensure the correct aim of the beam from such an enormous distance, and safety controls would quickly shut down the satellite transmitter in the event that the beam did wander off the target receiver. The design of the safety control system would actually be simpler than that of a nuclear power plant and the consequences of a failure would most likely be considerably less. Before such a plan is considered, however, intensive research should be undertaken to study the long-term effects of low-level microwave radiation; accidents are inevitable and portions of the population could occasionally be bathed in microwaves.

Another question is economic. The construction of the network as envisioned requires building dozens of enormous photovoltaic power satellites composed of tons of materials and millions of high-efficiency solar cells. The capital requirements are so large that only the most industrialized nations or a consortium of many giant corporations could even consider the project. The great number of twice-weekly space shuttle trips necessary to deliver the parts to synchronous orbit would consume an enormous amount of petroleum and could cause widespread environmental effects. After a three-year preliminary feasibility and design study, the federal government decided to suspend funds for further work on the solar power satellite because of these economic and environmental problems.

## A Proposed Alternative

Among other advantages, it may be safer and cheaper in the long run to start a space colony that has solar cell manufacturing facilities and utilize raw materials from the moon or asteroids to construct the power satellites. Such a proposal has been worked out in some detail by Gerard O'Neal of Princeton and others. Manufacturing solar cells in space may turn out to be a very profitable enterprise. Many of the problems that plague the manufacture of inexpensive, high-quality silicon cells may be simplified where a gravity-free environment and a high-quality vacuum are so easy to obtain.

Certainly the space colony would use solar energy to meet all its energy needs, and the plentiful supply of energy would also help the fledgling space solar cell industry. Solar cell manufacture is presently an energy-intensive business. In planning to alleviate this situation, calculations have been made of a manufacturing facility's growth rate if a given percentage of its output is dedicated to furnishing its own power needs. The only external supply of energy then needed would be that used to construct the initial nucleus of the factory; all other energy needs come from the dedicated output. In space, where AM0 (1.4 $kW/m^2$) of sunlight is continuously available, the growth rate for a solar breeder could be fast indeed.

Rather than build a solar power satellite, it might be more profitable yet for the space colony to ship manufactured solar modules to earth to be sold directly to homeowners as photovoltaic roof shingles. This approach would eliminate the cost and complexity involved in converting solar electricity to microwaves and constructing the receiving antennas and power conditioning equipment needed to convert the microwaves back to electricity. Instead of continuously beaming energy, the space colony would only have to ship a bundle of shingles once; the downhill trip on the space shuttle would then be made profitable, since it would otherwise return to earth empty.

The total area required for the proposed power satellite's receiving antennas is a significant fraction of the roof area of the customers of the electricity generated; actually it is not that much less than the area required by the terrestrial photovoltaic arrays needed to produce the equivalent amount of electricity.

The problems associated with this scenario are common to all decentralized electricity-generating networks. The perennial problems of balancing energy supply and energy consumption would be minimized by increasing the size of the utility grid and incorporating centralized short-term storage. The decentralized producer network could actually be made more reliable than our present patchwork utility grid. The social and political problems would be much more difficult to solve. Our society would have to change from one whose consumers depend completely on a few large corporations and government agencies for their energy needs to a society whose consumers are self-sufficient, but cooperating, producers of energy. Some large institutions are already resisting the tentative steps in this direction. However, these steps are being taken by pioneering individuals and groups who use sun, wind, and water to generate more electric power than they need and who sell the excess to local utilities. These pioneers have succeeded in showing the way to a new concept of utilities. The price of solar cells will come down, we will eventually have decentralized power grids, and photovoltaics will become the most practical way to furnish the energy our civilization requires.

## RECOMMENDED READINGS

Commission of the European Communities. *Sixth E. C. Photovoltaic Solar Energy Conference* (Dordrecht, Holland: D. Reidel Publishing Co., 1985).

Davidson, Joel. *The New Solar Electric Home: The Photovoltaics How-To Handbook* (Ann Arbor, MI: **aatec publications**, 1987).

DeMeo, E. A., and R. W. Taylor. "Solar Photovoltaic Power Systems: An Electric Utility R&D Perspective." *Science* 224(4646):245–51.

*Home Power Magazine*, PO Box 520, Ashland, OR 97520.

Maycock, Paul, and Edward Stirewalt. *Photovoltaics: Sunlight to Electricity in One Step* (Andover, MA: Brick House Publishing Co., 1981).

Photovoltaic Venture Analysis. Final Report. Solar Energy Research Institute, SERI/TR–52–040 (1978).

*Solar Today*, American Solar Energy Society, 2400 Central Avenue, G–1, Boulder CO 80301.

# Appendix A
# Assembling Your Own Solar Modules

It is possible for you to assemble your own photovoltaic modules, starting with individual cells. Although it is difficult to match the reliability and durability of the standard laminated commercial PV module, there are a number of reasons why this might be worthwhile. You might need a custom module for a special purpose that requires an unusual voltage or even multiple voltage outputs. Or you may need a module with a special shape to fit a particular location. Perhaps you have acquired a very inexpensive set of solar cells, maybe surplus or from an electronic "flea market," and you want to put them to work. Or you may simply want to learn more about solar cells and prefer the "hands-on" approach.

Over the years I have taught numerous workshops both in the U.S. and abroad where participants have assembled their own solar modules. Actually handling the solar cells helps to "demystify" the photovoltaic cell and to make it seem a more real, practical device. This provides a great learning experience for children also, and lately I have been working with the Maine Solar Energy Association (MSEA) and other groups to bring hands-on workshops into the classroom. MSEA also offers summer weekend workshops where participants assemble large hybrid modules which some participants purchase and take home. It is an enjoyable and inexpensive way to enjoy a weekend in Maine.

## TESTING SOLAR CELLS

There are a number of sources from which the average person may obtain individual solar cells. (Appendix B lists names and addresses of manufacturers; some, such as SunWatt and AstroPower, will sell to individuals.) These cells range from well-tested and calibrated devices guaranteed to perform to their given specifications through standard production-quality cells to untested or reject cells that can be purchased at a substantial discount. Even if you purchase a standard guaranteed cell, it is important to test each one thoroughly before using it. The amount of work entailed in removing one cell from a finished module and the danger of ruining the neighboring, good cells make testing individual cells a must.

Fortunately, testing solar cells is neither difficult nor complicated. The apparatus needed can be as simple as an inexpensive multimeter or voltmeter and ammeter of the proper range. Electronic supply houses sell these meters for $10 and up. A fancy digital meter is not needed, but it is good to have one that has among its selection of ranges a voltage range as low as 1 volt or less and a current range above 10 amps. To test completed modules, we will be measuring voltages of 16 to 18 volts or more and short-circuit currents possibly as high as 2.5 amps, depending on the size of the module and the light conditions.

The simplest test which gives meaningful information measures a cell's short-circuit current and open-circuit voltage under bright sunlight conditions. If bright sunlight is not available, it is possible to perform the tests outside or with the light through a window on a cloudy day or even under artificial illumination. To get meaningful results under these conditions, it is necessary to have a standard cell of known output available to measure the actual light intensity. Even then, if the light used in testing the cells is very different in color from that of real sunlight, any difference between the spectral response of the cells to be tested and that of the standard cell will produce an error. For example, on a cloudy day when the illumination is much bluer and "colder," a blue-sensitive cell will respond more efficiently than it would to "warm" incandescent illumination of the same intensity. However, if the standard cell is made of the same material as the cells to be tested, these differences will be negligible. What we aim for is not a set of 32 perfect cells, but rather a matched set of cells of roughly equivalent quality—even if that quality is relatively poor compared to the best cells available.

## A Simple Solar Simulator

To test solar cells indoors, you need a solar simulator. Usually, solar simulators are expensive devices with xenon lights and filters, but a very simple way to produce a reasonable simulation of AM1 and AM2 sunlight is to use a standard EHL projector lamp. These lamps, which are used in slide projectors, have special reflectors that are designed to let the infrared light escape from the bulb and that reflect only the visible light onto the slide to keep the slide cool. By coincidence, the spectral output of this lamp is very close to that of sunlight AM1 conditions if the lamp is operated at 117 AC. If the lamp voltage is reduced to 100 volts, the output becomes redder and nearly matches AM2 sunlight. The exact distance between the bulb and the solar cell will vary with each bulb, but it is about 35 cm. You can determine the exact distance with a solar cell that has been previously calibrated, but even without this the simulator will give a constant light output so that the cells can be checked relative to each other.

## A Simple Sample Holder

To measure the electrical properties of solar cells, it is necessary to make good electrical contacts to both the back contact and the top fingers in such a way that the cell is not shadowed. Figure A.1 shows a simple sample holder that can be built to do this. The base is a 6- x 8-inch piece of plywood onto which a piece of thin copper or galvanized steel is glued with contact cement. The five-way binding posts, available at most electronic supply stores, serve to connect the wires from the multimeter or voltmeter and ammeter to the holder. After drilling a pair of small holes 3/4 inch apart for the shafts of the binding posts, drill a larger hole from the back for each part so that

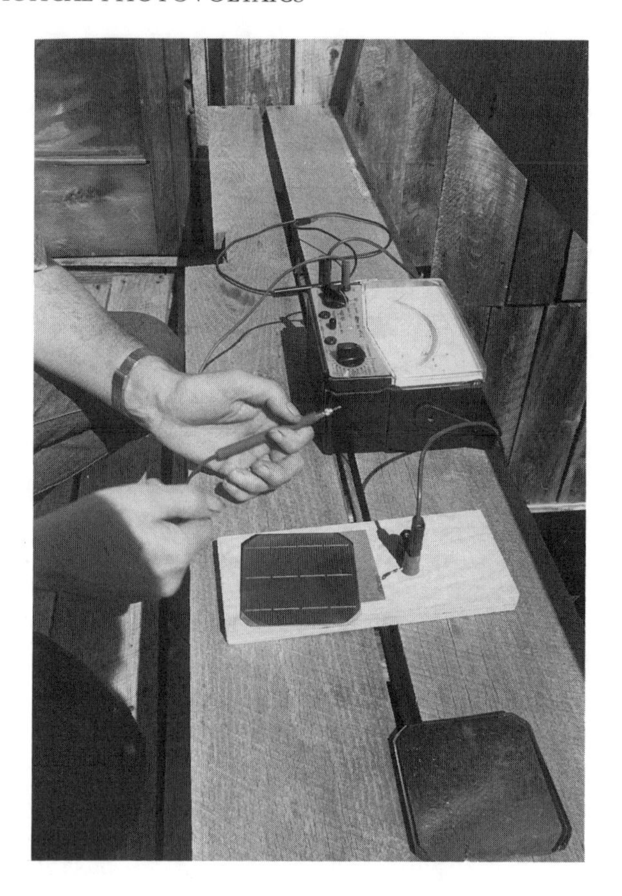

***Figure A.1.*** A simple sample holder. (David Ross Stevens photo)

the nuts holding the posts in place can be recessed, allowing the plywood base to sit flat. When a cell is to be tested, the back contact is established by resting the cell on the metal plate. (It is important that the plate be kept clean and free of insulating oxides.) The slight pressure of the front contact against the fingers should be enough to make a good back contact.

Two types of front contacts are possible for this holder. The simplest is just one of the probes that comes with a multimeter. It is the easiest to use if the holder is kept flat and there are a lot of cells to test: a cell can simply be laid on the holder plate and touched with the probe. A spring-style holder is easier to use for intensive testings of a single cell: it frees the hands and holds the cell securely in place so that the holder can be tipped to face the sun directly. Also, since the pressure is constant, variations in the measured short-circuit current caused by poor contact are minimized.

More elaborate sample holders with temperature-controlled bases and adjustable top contacts can be constructed for more advanced research into the properties of semiconductor solar cells. But for our testing purposes, the holder shown should be adequate.

## CURRENT–VOLTAGE MEASUREMENTS

All commercially available silicon solar cells produce nearly the same voltage. Even for relatively inefficient cells, the open-circuit voltage should be over 0.5 volt and not exceed 0.6 volt under direct sunlight illumination. This voltage will change only slightly with changes in the brightness of the sunlight. Under reduced illumination, the open-circuit voltage will drop to 0.3 volt or so.

The occasional defective cell with little or no voltage output probably is shorted. If you examine the cell and find no obvious defect, like a spot where the front contact finger has accidentally been extended to the edge so it can touch the back contact, the short is internal and cannot be fixed. If you are positive that the measurements have been made correctly and the short persists, it is possible to rescue part of the cell by deliberately breaking the cell in half, retesting, and using the good half. If the short is a point defect, the rest of the cell area will be perfectly good. Of course, the bad half can also be broken further, since each part of a solar cell—no matter how small—is a complete solar cell. With some clever jumper wires between the fingers, broken pieces of cells can be assembled into usable modules.

The current output of a solar cell is directly related to the area of the cell and the intensity of the light falling onto it. It is also dependent on the quality of the cell and should be the main determinant of whether or not to use a particular cell in a module, since the maximum current output of a series string of cells will be essentially the output of the worst cell in the string. The rated or AM1 output of a cell is the current expected from sunlight coming from directly overhead through a dry atmosphere (the Sahara Desert at high noon) and is the one most often used in literature and specifications. The AM2 output is closer to what would normally be expected from measurements made outdoors under normal weather conditions. For silicon cells not listed in the table, a figure of 27 mA/cm$^2$ for AM1 or 21 mA/cm$^2$ for AM2 conditions could be used. Simply calculate the surface area of the cells (including the fingers) in square centimeters and multiply by those figures to calculate the expected current output. If a standardized cell is available, it can be used to measure the exact light intensity at any given moment and will greatly simplify the checking process. You must be sure, though, that the standard cell and the cell to be tested are exposed to exactly the same light and are oriented at the same angle.

## HOW CELLS ARE CONNECTED

Solar cells can be connected to an external circuit in a number of ways. They can be fitted into a holder with spring contacts that press onto the front and back; the connections can be spot-welded; or the circuit they power can even be built onto the same chip of silicon as the cell area. But the most common way to connect solar cells is by soldering jumpers onto them.

Solar cells can be purchased with a current rating of 2.5 amps, but the half-volt output of a cell is usually too little for practical systems. The whole purpose of assembling cells into modules is to increase the voltage or current available compared to the individual cell output. Usually, the cell is sized to deliver the required current, and these cells are connected in series to produce the required voltage. Even if the current output desired is that of two or more individual cells, it is probably better to make these up as a set of two or more separate modules and then connect the modules to the external circuit. A blocking diode should always be used in connecting an array to the battery to keep current from flowing backward through the array at night or any other time when the array is putting out less than the battery voltage. If a separate blocking diode is put in series with each module in an array, the failure of one module will not affect the other. Figure A.2 shows several methods that have been used to build a series string of solar cells.

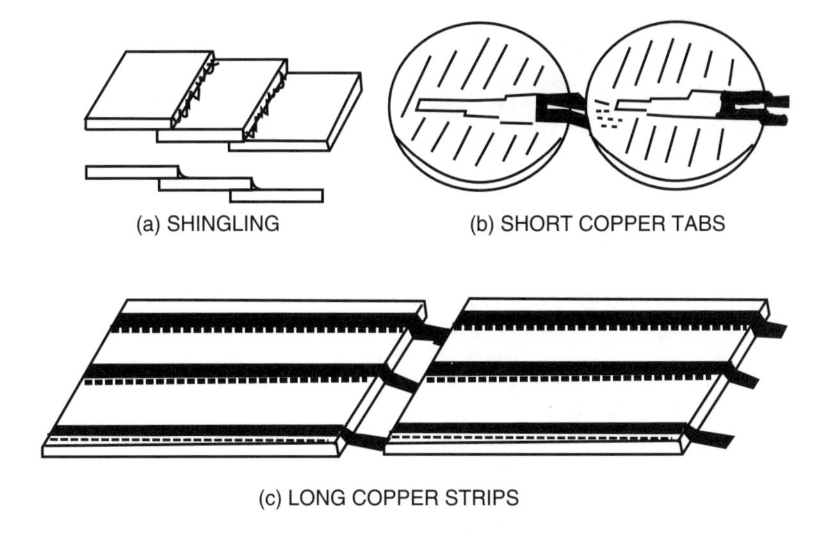


***Figure A.2.*** Several ways of connecting solar cells in series.

Early in the space program, when solar cells were used to power the first satellites, the usual practice for assembling a series string of cells was called **shingling** (shown in Figure A.2a). The bottom edge of one cell overlapped the top edge of the next cell by about 1 mm and was soldered. The finger pattern was designed with a soldering strip on one edge. However, expansion and contraction with temperature, along with any flexing, places an enormous stress on the solder junction and premature failure was one result.

At the present time, the most common way to interconnect solar cells is to use jumpers made of copper foil. Commercial solar modules have specially shaped foil pieces preplated with tin or solder, but straight strips can be used with most finger patterns.

It is quite easy to prepare jumper strips out of copper foil; the series connection methods are illustrated in Figure A.2 b & c. The copper foil should be quite thin (0.002 inch or 50 microns) in order to allow flexible connections that will put little stress on the cells themselves. These strips are very easy to cut from a wide piece of copper foil (see Figure A.3) or the foil used to assemble stained glass pieces can also be used. (This foil is available at many hobby shops or stained-glass supply dealers.) The exact size and shape of the strips will depend on the finger pattern of the cells, but it is best to

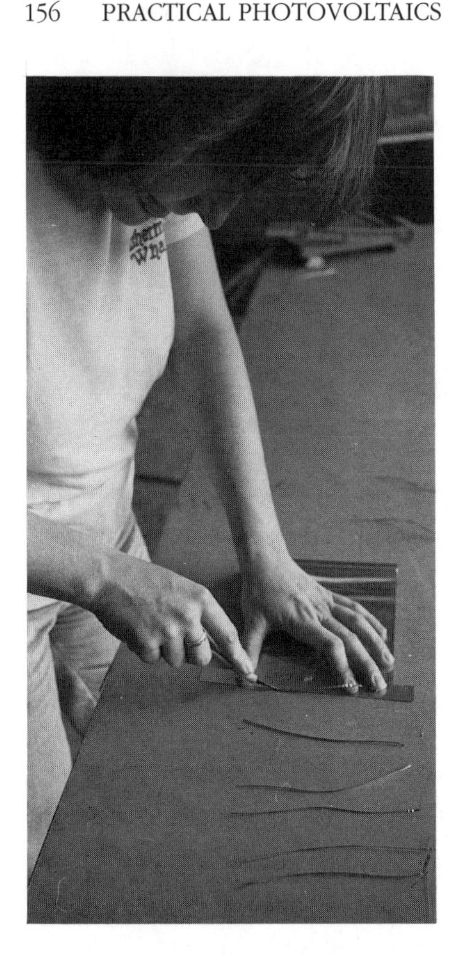

***Figure A.3.*** Preparing strips of foil. (David Ross Stevens photo)

allow about 1- to 2-mm space between cells. It is also advisable to arrange for two or three parallel strips between each cell to make redundant contacts. This way, the module will not stop working if one solder joint fails, as could happen with single jumper strips. An example of a cell where three jumper strips are possible is shown in Figure A.4. In this design, three wide bars connect all the fine fingers allowing multiple paths for current to flow around a gap or defective finger. It is recommended that thin, preferably crinkled, strips of copper be sweat-soldered the entire length of the wide bands. These strips should also run at least halfway across the back of the next cell. This will allow a cell that accidentally gets cracked to continue producing power at nearly full output.

Instead of copper foil, it is possible to use small-diameter copper wire as the interconnections. B&G No. 24 to No. 26 wire is the most useful size. At least two pieces of wire and preferably three or more should be used. One source for this thin-diameter wire is multistranded No. 16 or No. 18 wire. Once the insulation is stripped, you have 20 or more strands of wire prepared to the same length in one operation. The wire interconnections should be arranged so that they have a slight bend when the cells are positioned properly to allow for thermal expansion and contraction.

***Figure A.4.*** Soldering copper strips to a solar cell.
(David Ross Stevens photo)

## SOLDERING SOLAR CELLS

Most commercially available solar cells are designed to have contacts and wires soldered to them. The back contacts and front fingers are applied to the cells in a number of ways, including electrodeless nickel plating, vacuum metallizing and silk-screening. In addition, many manufacturers tin, or coat with solder, the contact points where jumpers are expected to be fastened. Ordinary electronic-grade (60% tin/40% lead) solder can be used in most cases, but a special solder which contains 2% silver in addition to 60% tin and 38% lead is the best choice if the cell contacts are silver. This will keep the hot solder from dissolving and possibly weakening the bond of the very thin silver fingers. Rosin core solder and rosin-based soldering flux can be used but can be difficult to clean off the cell faces once the module is soldered together. A water-soluble flux is available (Kester Solder Co., Chicago, Illinois) that is normally used by solar cell manufacturers. This flux simplifies cell cleaning after the interconnections have been soldered.

A small pencil-type soldering iron of about 35 to 40 watts capacity is the most useful for this kind of soldering. The cone shaped (pencil) tip or the small screwdriver-shaped tips seem best able to deliver a precise amount of heat to a small area.

The actual soldering of the copper foil pieces to the cells is not too difficult but it does require fast work with a hot iron. As with any soldering job, it is most important that the surfaces are clean. It is best to clean solar cell surfaces with a cotton swab and isopropanol (rubbing alcohol) and let the surface dry before soldering. Some solar cells, like those from Solarex, have an antireflection coating covering the entire top surface, including the fingers. To be able to solder to these cells, first clean the spot on the finger by "erasing" (rubbing lightly with a pencil eraser) until the spot looks shiny and metallic. Then tin the area with ordinary 60/40 rosin-core solder and solder the ribbon to the tinned spot. No extra solder or paste is require for this step. If the cell contact is already tinned, put a thin layer of soldering paste on the copper foil and sweat-solder the foil with the soldering iron. Figure A.4 shows this operation. Use firm pressure, but not enough to crack the cell. No solder is necessary. If the cell contact is not tinned, tin the foil strip first. The strip must be tinned on every surface that will be soldered, but try to leave an untinned place where the foil jumps from one cell to the next, in order to increase the flexibility. It is best to solder all the foil tabs to the finger side of each cell first, then arrange the cells face down in the position desired for the module. Taping the cells face down to a piece of hardboard in the desired pattern will make it easy to align the cells properly. Then solder down the tab to the back of each cell. If small pieces of masking tape are used, they can be removed easily without breaking the cells. It is best to make the contacts to the final cells in the module with the same kind of foil and then solder wires to the foil a centimeter or so away from the cell to reduce the strain.

If wires are used as the jumpers, they are soldered in a manner similar to that used for the foil strips. The only difference is that it is generally not necessary to pre-tin the wires before fastening them to the cells.

## BUILDING A FLAT PLATE MODULE

One simple way to offer protection for the cells is to place them in a flat acrylic box. The collector box has a removable lid that is fastened down with screws and sealed with silicone grease. Though it was not sealed tightly enough to withstand constant exposure to the salt spray on the deck of an ocean-going yacht, a module installed in such a box has been operating outdoors for fourteen years without any decrease in electrical output.

### Laying Out the Case

To construct the flat plastic module, the first step is to lay out the arrangement desired for the solar cells. For a 32-cell module, I usually use 4 rows of 8 cells each, but other arrangements are possible: 3 rows of 11 cells for 33 cells, 6 rows of 6 cells for a square 36-cell module, or 4 rows of 9 cells if a rectangular 36-cell module is desired. Other combinations are possible with unequal row lengths—for example, 3 rows of 6 plus 2 rows of 7 for an almost square 32-cell module. I prefer to put the cells in straight rows separated by thin acrylic strips. The strips keep the rows of cells separated without fastening the cells to the case, and they also help support the lid of the case so that it cannot be pushed inward enough to press onto and possibly crack the cells.

Once an arrangement has been devised, you can calculate how big the case should be. The easiest way to do this is to actually fasten the cells into their final arrangement, lay them onto a flat surface, and measure the set, allowing about 1/2 inch between each row and another 1/2 inch at each edge. Figure A.5 shows a case built for 3-inch circular cells. The dimensions given in the drawing will allow plenty of room for expansion and contraction. They also allow for some error in soldering the cells into absolutely straight rows, but at the same time prevent the sets of cells from moving too far in relation to each other and the case.

## Cutting the Plastic

Acrylic sheets (also called Plexiglas™, the tradename for Rohm and Haas plastic) come in a variety of thicknesses and qualities. Cast acrylic sheets are a little easier to use and glue, but they are more expensive; the extruded sheets are cheaper and seem to be a bit stronger. The extruded sheets are 1/10-inch thick (called 100 mil by

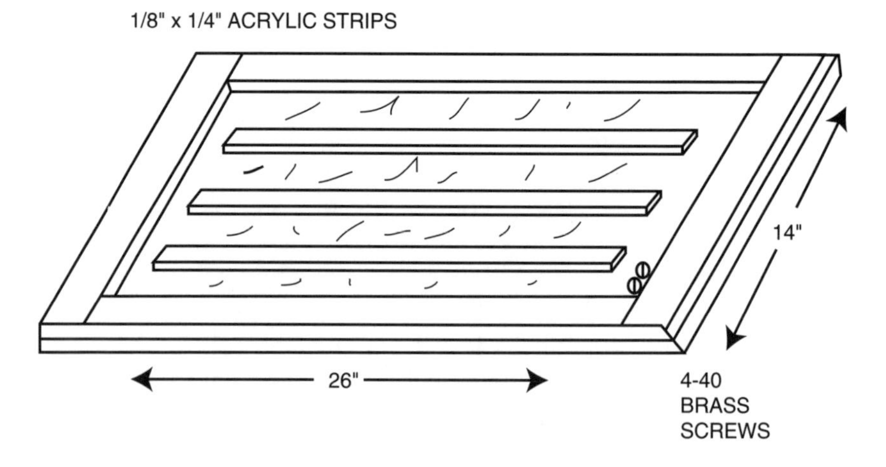


***Figure A.5.*** Drawing of a plastic case for a flat plate module. The dimensions shown are for 75-mm (3-inch) circular cells.

glaziers) and are commonly used to glaze storm doors when using glass would be too dangerous. This thickness is fine for solar cell modules of the size discussed here. With cast acrylic, I generally use the 1/8-inch thickness, which is more than adequate for most purposes. If the module case will be subjected to a great deal of abuse, 3/16- or even 1/4-inch could be used for the front and back covers, but you should use 1/8 inch for the spacer strips unless you plan to put some sort of backing material behind the cells to cushion them. For even more strength, polycarbonate sheets can be used instead of acrylic.

Cutting acrylic is quite simple. The easiest way to cut the 100 mil sheet is to score it with a utility knife and then break it along the score by bending it away from the scored side. Use a metal straight edge to get a straight scratch and score the sheet as deeply as possible. Unlike glass, acrylic can be scored by several passes over the same line, and if the knife cuts almost all the way through in spots, so much the better. Cutting 1/4-inch-wide spacer strips by this method may be difficult unless you are proficient at working with the material.

Acrylic can also be cut with a hand saw or a saber saw or even a table saw or portable circular saw. The problems here are chipping or cracking the plastic if too coarse a blade is used, or melting the plastic if too fine a blade is used. I have cut acrylic only to find the molten plastic and sawdust welded back together behind the saw blade. When using a circular saw, a veneer cutting blade seems to be the best compromise between melting and chipping.

Polycarbonate is very similar to acrylic when cut with a saw, but it cannot be scored and broken as easily.

### Gluing the Plastic Case

There are a number of commercially available glues for acrylics. Most of these work on a solvent principle: they dissolve the acrylic slightly to make a sticky surface. I usually use a chlorinated hydrocarbon such as trichloroethylene or chloroform. Chloroform is probably the best solvent glue for acrylic, but it is extremely dangerous and is suspected of causing cancer. Avoid breathing the fumes or getting it on your skin. The commercial glues, while still very dangerous, do not require the extreme caution necessary in handling chloroform.

The best solvent glue for polycarbonate is acetone, but methyl ethyl ketone will also work.

To use solvent glue, place the pieces to be joined in the desired position and, using an eyedropper, let a few drops of the solvent run into the crack between the pieces. Capillary action will pull the solvent into place. Light but steady pressure on the joint will produce a complete bond. Do not move the pieces for five minutes, then examine the joint to see if it has been glued evenly. The glued parts will look wet and transparent when viewed through one of the shiny faces of the pieces, while the spots that are not actually in contact will reflect more light. You can apply a few more drops of solvent to these spots to glue them. It is possible, using this technique, to make a completely sealed, waterproof joint as strong as the plastic.

### Fastening the Cover

When you have all the edge and divider strips glued onto the backing sheet, you are ready to install the solar cells and fasten down the cover. If you are sure that the module is soldered correctly and working properly, thus needing no maintenance for quite some time, you can attach the cover using the procedure just described. But a better idea is to fasten the top cover with screws.

After the plastic case is glued together, but before you install the cells, lay the top cover temporarily in place and drill and tap a set of holes around the entire outer edge of the case. I usually fasten the cover with 440 brass-headed screws. To ensure that all the holes will align, install two at opposite ends before drilling the rest of the holes. Drill and tap a couple of holes into the center divider strip to help hold the case together.

### Installing the Solar Cells

Now that the cover is ready, remove it and lay the strips of cells into the case. Sheets of polyethylene-foam packing material (1/16-inch thick) can be fitted into the bottom of the plastic case as a cushion behind the solar cells. (Do not use foam rubber for this purpose because it deteriorates in sunlight.) You can also fasten the cells to the case back with a dab or two of silicone RTV behind each cell. This produces a cushioned, permanent mounting; in addition, the soft RTV will absorb any thermal expansion movement. However, once the silicone has cured, the cells are impossible to remove without breaking; make sure that all the connections are good before taking this irreversible step. If you turn every other string end for end, the jumper wire from the back of the last cell in one string will be at the same end as the jumper wire from the front of the first cell in the next string, making the connections between strings short and easy to make. It is best to slip a piece of cardboard under the jumpers when soldering them in the case because acrylic cannot withstand much heat.

There are several methods for bringing the final wires out of the case. The easiest is to simply file a couple of notches into the top of one edge strip, one on each side of a cover screw, and lay the wires in these notches so that the cover clamps them in place. However, the danger of this arrangement is that someone might yank on the wires and break a cell where it is fastened to the wires. A neater solution is to attach a pair of binding posts to the back, for example, a pair of 6–32 screws that are screwed from the inside through holes tapped through the rear cover. Use a little silicone caulk under each screw to seal the joint. I make the holes exactly 3/4 inch apart so a dual banana jack can be plugged onto the connection if desired. If it is small enough, you can attach a blocking diode inside the case to one of the binding posts. The cathode end of the diode (the end with the white band) goes to the positive (+) binding post while the anode end is fastened to the rear of the last cell in the string. Of course, the front of the first cell is connected to the negative (-) binding post.

After you have assembled and wired the module, try it out in the direct sunlight or in as bright a light as possible. Then clean the inside of the front cover and install it, using some silicone grease all around the edge as a sealant. Clear silicone caulk can also be used, as it does not adhere very well to the plastic and it will be possible to pry up the cover if you have to get inside to repair the module.

## BUILDING A HYBRID CONCENTRATOR MODULE

Figure A.6 shows a second type of module made at a Skyheat workshop. This hybrid module has two advantages over the flat module described above: (1) it can use a Winston concentrator to increase

***Figure A.6.*** Skyheat workshop-fabricated hybrid solar cell array installed on a greenhouse. (David Ross Stevens photo)

the amount of sunlight striking each cell, and (2) it has provisions for heating either water or air with the waste heat produced in the solar cells. We have made two slightly different versions of this collector: one with a sheet metal backing support and water tubes; one with the cells mounted on a strip of wood. The second version can be used only for heating air, but it is simpler and cheaper to make. Both versions will be described here.

## The Backing Support

The wooden backing support is simply a piece of wood about 1/2 inch wider than the cells and long enough to carry the desired string of cells. It is probably a good idea to limit the length of a single board to 4 or 5 feet, making the string of cells 16 cells long. Two of these modules then make up one module. If the backing support is made exactly 4 feet long, it is possible to fix 16 three-inch cells onto it. Since each cell has a small flat on one end, the cells can be mounted slightly less than 3 inches apart, leaving a millimeter or so between cells.

A backing support bent from sheet metal will conduct heat much better than wood. A couple of small tubes soldered to it serve to carry water to cool the cells and deliver the heat to a hot water heater. Figure A.7 is a drawing of one possible arrangement.

It is best to make the sheet metal support and the cooling tubes out of the same material to keep differential thermal expansion from warping the system, but small (1/4-inch–o.d.) copper tubing can be soldered into the corners of the backing plate and the combination does not warp. The dimensions shown in Figure A.7 are for a concentrator system using 3-inch–diameter cells. Larger or smaller cells can be accommodated by changing the width, but the 1-inch depth of the legs is good for stiffness. 18-gauge galvanized sheet metal is thick enough for this use.

Another possible backing support for this kind of concentrator module is an aluminum extension. Manufacturers of solar hot water collectors make an extension 6 inches wide and flat on the top surface with a groove in the back that is designed to be fitted with a

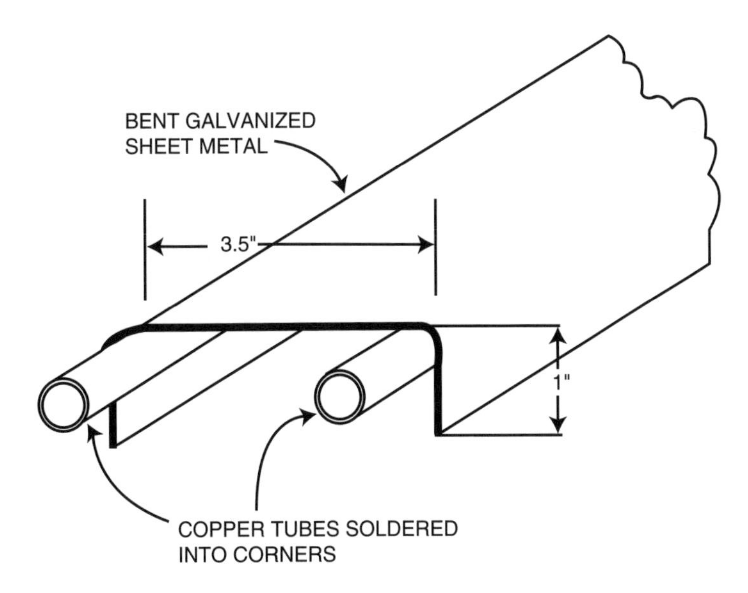


***Figure A.7.*** Sheet metal backing support.

piece of 5/8-inch–o.d. copper tubing. This makes a rather simple and elegant holder for solar cells. Modules with this type of backing support have been successfully constructed at several workshops.

## Isolating and Encapsulating the Cells

If a metal backing support is used, the string of cells must not touch the backing because that would short them. Therefore, the cells must be fastened to the backing so that electrical insulation and adequate heat conduction are ensured. Silicone RTV (transparent silicone rubber caulk) is a very good electrical insulator, a moderately good conductor of heat, and has the added advantage of adhering extremely well to silicon cells. Based on the use of this material, we developed the following method to fasten and seal cells. Two-part, transparent silicone RTV, such as General Electric RTV 615, can also be used with the following instructions. It is easier to use, but more expensive.

***Step 1.*** Using a caulking gun, lay a bead of silicone RTV down the center of the backing plate, and quickly spread this layer to cover the entire plate about 1/16-inch thick (Figure A.8). Speed is more important than perfection at this point.

***Step 2.*** Cut a strip of cloth about 1 inch wider and larger than the backing plate. (Prepare this strip in advance so that no time is wasted while the silicone is setting up. Use polyester or some other synthetic cloth so any moisture that might get into the layer will not rot the fabric.) Press the cloth onto the fresh silicone layer (Figure A.9).

***Step 3.*** Coat the cloth with another silicone layer, taking more care this time to get a level surface (Figure A.10). The purpose of these two steps is to create a barrier to keep the cells from accidentally touching the backing metal. For wood backs, Steps 2 and 3 can be eliminated if desired, although the cloth does strengthen and stabilize the silicone layer.

***Step 4.*** Lay the solar cells carefully onto the wet silicone layer (Figure A.11). It is best, if possible, to solder the copper jumper strips to the back of each cell and place the cells one at a time onto the backing. Then solder them together after testing each cell for proper operation and shorts to the metal base. Press each cell very gently into the silicone to work the air bubbles out from the back. (A few air

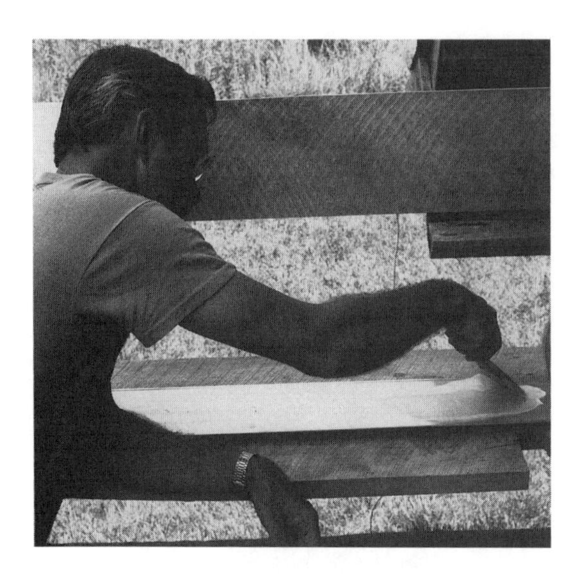

***Figure A.8.*** Spreading silicone RTV caulk onto the backing plate.

***Figure A.9.*** Placing the cloth onto the fresh silicone. (Skyheat Workshop photo)

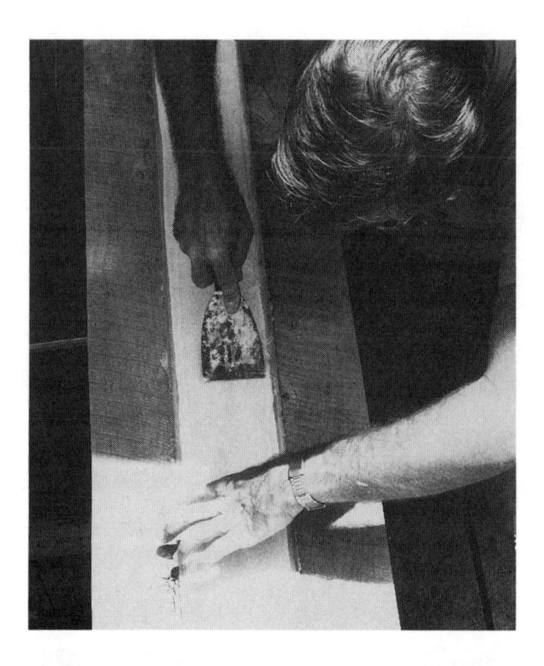

***Figure A.10.*** Smoothing out a new layer of silicone over the cloth, completely embedding the cloth. (Skyheat Workshop photo)

bubbles are better than a cracked cell, however.) If you must solder the string together before embedding, have someone hold up one end of the string while you carefully place each cell in turn. It is possible, if you are quick enough, to remove a broken cell and replace it with a good one while the silicone is still tacky. It is almost impossible to remove a cell once the module is finished.

***Step 5.*** When you are completely satisfied that the module is working properly, apply a top layer of silicone RTV over the entire module. I usually finish by rolling on a Mylar™ top film while the silicone is fresh. (3–M sells rolls of Mylar™ plastic as an accessory to their overhead projectors.) Simply place a bead of caulk on the cells and slowly roll the film over it, rolling-pin fashion, to spread the silicone (Figure A.12). With practice, this will produce a thin, uniform top surface without any air bubbles trapped atop the cells. Make sure that the wires projecting from the ends of the module do not strain the connections to the end solar cells. I usually seal a couple of S curves in these wires so they are held in place by the silicone caulk.

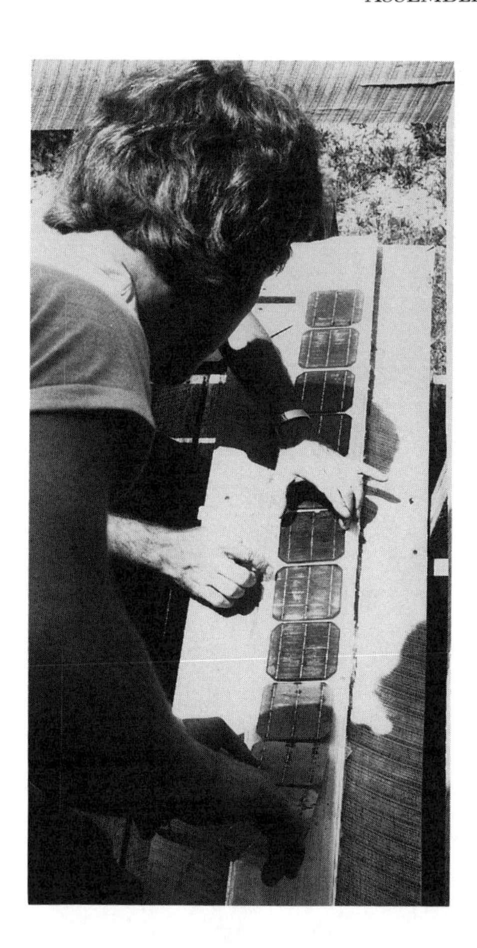

***Figure A.11.*** Laying the string of solar cells onto the freshly prepared backing. (Skyheat Workshop photo)

If it ever becomes necessary to repair a cell connection, it is possible to cut away the top encapsulant in the area to be repaired (a razor blade works well). Once the repair is made, a dab of silicone RTV smeared over the spot will reseal the system. If you are not sure where the defective spot is, you can use a multimeter with one probe replaced with a needle to poke carefully through the encapsulant to touch the jumper strips. Connect the other probe to one of the wires coming out of the module, and probe each jumper in turn. (Carefully confine this poking to the copper jumper strips because you can break a solar cell by hitting it too hard with the needle.) The bad jumper will be either the one which shows no reading, or the one next to it, depending on which end of the jumper has the bad con-

***Figure A.12***. Coating the top of the cells with silicone and Mylar film. The film presses out a bead of silicone RTV, eliminating voids and air bubbles. (Remote Power, Fort Collins, CO)

***Figure A.13.*** The finished product is a completely encapsulated set of solar cells. (Remote Power, Fort Collins, CO)

nection. I prefer to measure the short-circuit current with a meter when doing this kind of troubleshooting, as a bad solder joint may be leaky enough to allow the correct open-circuit voltage to be read, even though it will let virtually no current pass.

## The Reflector Wings

### *Design*

Once the long modules are assembled and sealed, they are ready to use just as they are, but much more effective use can be made of these expensive solar cells if reflectors are added. For low concentration ratios, the design is not critical. Figure A.14 shows the dimensions of several reflectors that can be used with this type of module. The first is simply a pair of tilted flat reflectors that will almost double the amount of sunlight falling on the cells. This simple system, while very easy to build, does not have the acceptance angle of the second design; thus, the tilt would have to be changed about four times a year for most efficient operation. The second system also has a concentration ratio of 2, but the curved surfaces are more effective in reflecting off-axis light onto the cells; the tilt angle need be changed only twice a year. This design works with reasonable efficiency even if set at a simple fixed angle (equal to the latitude from the vertical).

The third system is simply a 4-to-1 concentration ratio version of the same Winston design. These reflectors are quite tall and will need some bracing to keep their shape. Also, the exact shape is more critical. Wood-backed modules are not recommended for use with this high a concentration ratio: the power input to the cells is quite high and the stagnation temperatures reached on a sunny day will decompose the wood and could cause a fire. In fact, wood-backed collectors could be dangerous in any hybrid system where the chamber above the wood is sealed off from the outside air.

### *Making the Reflector Wings*

The easiest material from which to form the reflector wings is aluminum flashing. It is easily obtained at any lumber yard, stiff enough to hold its shape, and reasonably shiny. However, aluminized Mylar is much more reflective. The extra time and trouble taken to cover the

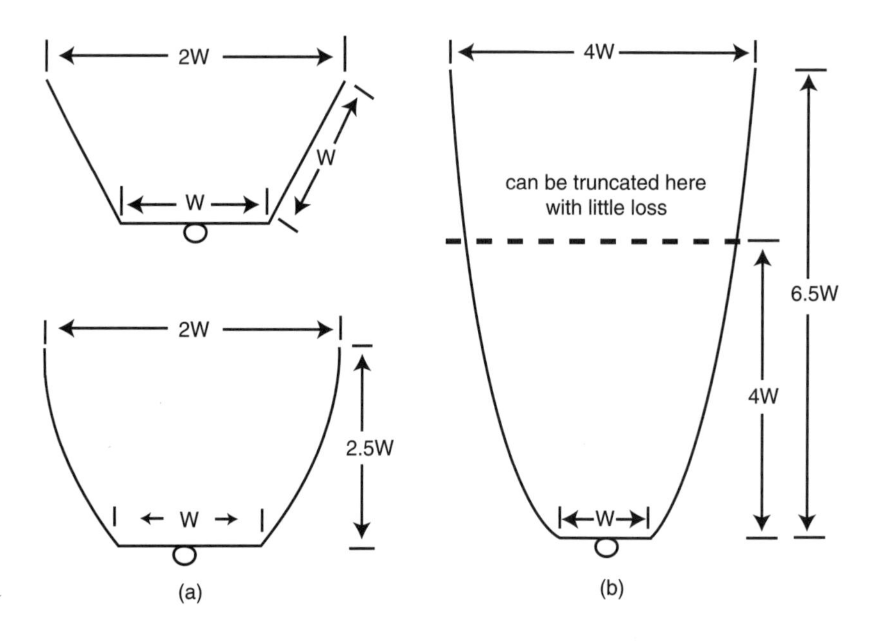


***Figure A.14.*** Dimensions of several reflectors: (a) 2-to-1 concentration ratio; (b) 4-to-1 concentration ratios.

reflectors with this material will make them considerably more effective, producing as much as 30% more electricity than a system with plain aluminum reflectors. Some aluminized plastics are designed for exterior use and will keep their high reflectance for years.

When bending the aluminum reflectors to shape, first put a sharp kink where the reflector is fastened to the backing strip. You can improvise a long bending brake for this light-gauge material by clamping the aluminum between two boards with C-clamps and bending the protruding edge by hand or with a third board. The gentle parabolic curve in the metal can be approximated by bending the sheet around a round object. (We use a piece of 3-inch plastic sewer pipe.) The greatest bend is closest to the crease, and the springiness of the aluminum will keep the outer edge to a barely discernible curve.

The wings can be fastened to the module backing with sheet metal screws. (Be careful not to drill through a solar cell.) Then make the final adjustments to the wing shape and angle. Prop up the module so that you can look into it from about 15 feet away. You will see a distorted reflection of the solar cells in each wing if you look at the module from straight on-axis. Bend and adjust the reflectors until the blue solar cell images seem to fill the reflectors completely. By moving your head to one side, you can get an idea of how much off-axis sun angle the reflector system will accept.

If two or more modules are installed side by side, they can be mounted so that the top edges of the reflectors just touch. The reflectors can then be fastened together, producing a very rigid support.

## A Hybrid Hot Air System

Figure A.15 shows two ways of ducting air through the collector to make a hybrid hot air system. The first incorporates a transparent cover over the entire module to create a series of plenum chambers out of the spaces between the reflector wings. The cover is not a bad idea in any case as it will further protect the cells from the weather and keep the reflectors clean, at a slight loss in efficiency due to the reflection losses in the cover material. If the air is blown through this chamber, the cover should be double-glazed. In the second ducting method, the air is static above the cells and blown through the space behind them. Single glazing is sufficient if this method is used. The basic system is amenable to many variations, and it will be interesting to see what clever innovations develop.

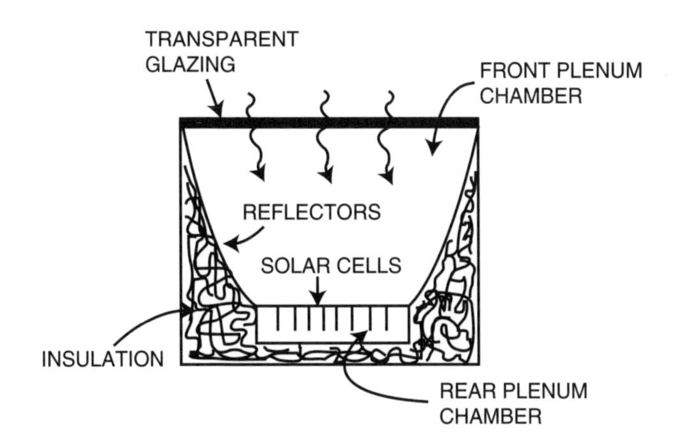


***Figure A.15.*** Two ways to duct air through a hybrid system: through either the front or rear plenum chamber.

# Appendix B
# Manufacturers of Solar Cells and Modules

ASE Americas
4 Suburban Park Drive
Billerica, Massachusetts 01821
(978) 667-5900
www.asepv.com

AstroPower, Inc.
Solar Park
Newark, Delaware 19716-2000
(302) 366-0400
www.astropower.com

Atlantis Energy, Inc.
4610 Northgate Boulevard, #150
Sacramento, California 95834
(916) 920-9500
www.atlantisenergy.com

BP Solar
P.O. Box 4587
Houston, Texas 77210-4587
(707) 428-7800
www.bpsolar.com

Carrizo Solar Corporation
1320 12th Street NW
Albuquerque, New Mexico 87104
(505) 764-0345

Crystal Systems, Inc.
27 Congress Street
Salem, Massachusetts, 01970-5597
(508) 745-0088
www.crystalsystems.com

Ebara Solar, Inc
811 Route 51 South
Large, Pennsylvania 15025
(412) 382-1251

Energy Conversion Devices
1675 West Maple Road
Troy, Michigan 48084
(810) 280-1900
http://ovonic.com

Energy Photovoltaics, Inc.
P.O. Box 7456
Princeton, New Jersey 08543
(609) 587-3000

ENTECH, Inc.
1077 Chisolm Trail
Keller, Texas 76248
(817) 379-0100
www.entechsolar.com

Evergreen Solar, Inc.
211 Second Avenue
Waltham, Massachusetts 02451
(781) 890-7117

Global Solar Energy, L.L.C.
12407 West 49th Avenue
Wheat Ridge, Colorado 80033
(303) 420-1141

Hoxan America Inc.
One Centennial Plaza, #3F
Piscataway, New Jersey 08854
(908) 980-0777

Internat'l Solar Electric Technology, Inc.
8635 Aviation Boulevard
Inglewood, California 90301
(310) 216-1423

Iowa Thin Film Technologies
2501 North Loop Drive
Ames, Iowa 50010
(515) 294-7005

Keep It Simple Systems
32 South Ewing, Suite 330
Helena, Montana 59601
(800) 327-6882
http://wildwestweb.com

Kyocera America, Inc.
8611 Balboa Avenue
San Diego, California 92123
(619) 576-2647
www.kyocera.com

Materials Research Group
12441 West 49th Avenue
Wheat Ridge, Colorado 80033
(303) 45-6688

Midway Labs Inc.
350 North Ogden
Chicago, Illinois 60607
(312) 432-1796
www.megsinet.com/midway

NEPCCO Environmental Systems
2140-100 NE 36th Avenue
Ocala, Florida 34470
(352) 867-7482
www.nepcco.com/enviro

Photocomm, Inc.
7681 East Gray Road
Scottsdale, Arizona 85260-3469
(800) 223-9580
www.photocomm.com

Photovoltaics International, LLC
3500 Thomas Road, Suite E
Santa Clara, California 95054
(408) 986-9231

Siemens Solar Industries
4650 Adohr Lane
Camarillo, California 93010
(805) 482-6800
www.solarpv.com

Solar Cells, Inc.
1702 North Westwood Avenue
Toledo, Ohio 43607
(419) 534-3377

Solardyne Corporation
20 South Main Street
Gainesville, Florida 32601
(352) 372-0333

Solarex Corporation
630 Solarex Court
Frederick, Maryland 21701
Phone: (301) 698-4200
www.solarex.com

Solec International Inc.
12533 Chadron Avenue
Hawthorne, California 90250
(310) 970-0065

Spectrolab, Inc.
12500 Gladstone Avenue
Sylmar, California 91342
(818) 365-4611

Spire Corporation
One Patriots Parks
Bedford, Massachusetts 01730-2369
(781) 275-6000
www.spirecorp.com

SunCat Solar
17626 North 33rd Place
Phoenix, Arizona 85032
(602) 404-4929
www.users.uswest.net/~tkreider

SunWize Energy Systems, Inc.
9415-19 Enterprise Drive
Mokema, Illinois 60448
(708) 479-1600
www.sunwize.com

TECSTAR Inc.
15251 Don Julian Road
City of Industry, California 91745-1002
Phone: (818) 968-6581
www.appliedsolar.com

United Solar Systems Corp.
1100 West Maple Road
Troy, Michigan 48084
(313) 362-4170

Utility Power Group, SECO Company
9410-G DeSoto Avemue
Chatsworth, California 91311
Phone: (818) 700-1995

For updates, and links to more information, see the Department of Energy PV website: www.nrel.gov/ncpv. Other useful sites include: www.solaraccess.com; www.homepower.com; www.crest.org; www.ases.org; and www.ises.org.

# Appendix C
# Current Carrying Capacity of Copper Wire

The ratings in the following table are those permitted by the National Electrical Code for flexible cords and for interior wiring of houses, hotels, office buildings, industrial plants, and other buildings.

The values are for copper wire. For aluminum wire, the allowable carrying capacities are taken as 84% of those given for the respective sizes of copper wire with the same covering.

| Size A.W.G. | Area Circular (mils) | Diameter of Solid Wires (mils) | Rubber Insulation (amps) | Varnished Cambric Insulation (amps) | Other Insulations and Bare Conductors (amps) |
|---|---|---|---|---|---|
| 24 | 404 | 20.1 | — | — | 1.5 |
| 22 | 642 | 25.3 | — | — | 2.5 |
| 20 | 1,022 | 32.0 | — | — | 4 |
| 18 | 1,624 | 40.3 | 3* | — | 6** |
| 16 | 2,583 | 50.8 | 6* | — | 10** |
| 14 | 4,107 | 64.1 | 15 | 18 | 20 |
| 12 | 6,530 | 80.8 | 20 | 25 | 30 |
| 10 | 10,380 | 101.9 | 25 | 30 | 35 |
| 8 | 16,510 | 128.5 | 35 | 40 | 50 |
| 6 | 26,250 | 162.0 | 50 | 60 | 70 |
| 5 | 33,100 | 181.9 | 55 | 65 | 80 |
| 4 | 41,740 | 204.3 | 70 | 85 | 90 |
| 3 | 52,630 | 229.4 | 80 | 95 | 100 |
| 2 | 66,370 | 257.6 | 90 | 110 | 125 |

Note: 1 mil = 0.001 inch.

*The allowable carrying capacities of No. 18 and 16 are 5 and 7 amperes, respectively, when in flexible cord.

**The allowable carrying capacities of No. 18 and 16 are 10 and 15 amperes, respectively, when in cords for portable heaters. Types AFS, AFSI, HC, HPD, and HSJ.

# Appendix D
# Conversion Factors

| To Change | Into | Multiply by |
|---|---|---|
| BTU | cal | 252 |
| BTU | joules | 1,055 |
| BTU | kcal | 0.252 |
| BTU | kWh | $2.93 \times 10^{-4}$ |
| BTU $ft^{-2}$ | langleys (cal $cm^{-2}$) | 0.271 |
| cal | BTU | $3.97 \times 10^{-5}$ |
| cal | ft-lb | 3.09 |
| cal | joules | 4.184 |
| cal | kcal | 0.001 |
| cal $min^{-1}$ | watts | 0.0698 |
| cm | inches | 0.394 |
| cc or $cm^3$ | in.$^3$ | 0.0610 |
| $ft^3$ | liters | 28.3 |
| in.$^3$ | cc or $cm^3$ | 16.4 |
| ft | m | 0.305 |
| ft-lb | cal | 0.324 |
| ft-lb | joules | 1.36 |
| ft-lb | kg-m | 0.138 |
| ft-lb | kWh | $3.77 \times 10^{-7}$ |
| gal | liters | 3.79 |
| hp | kW | 0.745 |
| inches | cm | 2.54 |
| joules | BTU | $9.48 \times 10^{-4}$ |
| joules | cal | 0.239 |
| joules | ft-lb | 0.738 |
| kcal | BTU | 3.97 |
| kcal | cal | 1,000 |
| kcal $min^{-1}$ | kW | 0.0698 |
| kg | lb | 2.20 |
| kg-m | ft-lb | 7.23 |

| To Change | Into | Multiply by |
|---|---|---|
| kW | hp | 1.34 |
| kW | kcal $min^{-1}$ | 14.3 |
| kWh | BTU | 3,413 |
| kWh | ft-lb | $2.66 \times 10^6$ |
| langleys (cal $cm^2$) | BTU $ft^{-2}$ | 3.69 |
| langleys $min^{-1}$ (cal $cm^{-2}$ $min^{-1}$) | watts $cm^{-2}$ | 0.0698 |
| liters | gal | 0.264 |
| liters | qt | 1.06 |
| m | ft | 3.28 |
| lb | kg | 0.454 |
| qt | liters | 0.946 |
| $cm^2$ | $ft^2$ | 0.00108 |
| $cm^2$ | $in.^2$ | 0.155 |
| $ft^2$ | $m^2$ | 0.0929 |
| $m^2$ | $ft^2$ | 10.8 |
| watts $cm^{-2}$ | langleys $min^{-1}$ (cal $cm^2$) | 14.3 |

# Glossary

**AC**—Alternating current; the electric current which reverses its direction of flow. 60 cycles per second is the standard current used by utilities in the U.S.

**acceptance angle**—The total range of sun positions from which sunlight can be collected by a system.

**acceptor levels**—Levels capable of accepting an electron from the valence band.

**AH**—*see* **ampere-hours**

**air mass 0 (AM0)**—The amount of sunlight falling on a surface in outer space just outside the earth's atmosphere (1.4 $kW/m^2$).

**air mass 1 (AM1)**—The amount of sunlight falling on the earth at sea level when the sun is shining straight down through a dry clean atmosphere. (A close approximation is the Sahara Desert at high noon.) The sunlight intensity is very close to 1 kilowatt per square meter (1 $kW/m^2$).

**air mass 2 (AM2)**—A closer approximation to usual sunlight conditions; may be simulated by an ELH projector bulb. The illumination is 800 $W/m^2$.

**alternating current**—*see* **AC**

**amorphous semiconductors**—*see* **semiconductors, amorphous**

**amorphous silicon**—A form of silicon with no long-range crystalline order used to make solar cells.

**ampere-hours (AH)**—A current of one ampere running for one hour.

**balance-of-systems (BOS)**—All the equipment needed to make a complete working photovoltaic power system—except the solar cell modules.

**band, conduction**—*see* **conduction band**

**band, valence**—*see* **valence band**

**band gap energy**—The energy needed to raise an electron from the top of the valence band to the bottom of the conduction band.

**band model**—The quantum mechanical model of solids that explains the behavior of semiconductors.

**black-body radiation**—*see* **radiation, black-body**

**battery, marine**—A deep-discharge battery used on boats; capable of discharging small amounts of electricity over a long period of time.

**battery, motive-power**—A large-capacity deep-discharge battery designed for long life when used in electric vehicles.

**battery, secondary**—*see* **battery, storage**

**battery, stationary**—For use in emergency standby power systems, a battery with long life but poor deep-discharge capabilities.

**battery, storage**—A secondary battery; rechargeable electric storage unit that operates on the principle of changing electrical energy into chemical energy by means of a reversible chemical reaction. The lead–acid automobile battery is the most familiar of this type.

**blocking diode**—A device that prevents the current from running backward through an array, thereby draining the storage battery.

**built-in potential**—The electrical potential that develops across the junction of two dissimilar solids. The open-circuit voltage of a photovoltaic cell approaches, but is always less than, this potential.

**bulk recombination**—*see* **recombination, bulk**

**carriers, majority**—The carrier most present in a doped semiconductor: holes in p-type, electrons in n-type.

**carriers, minority**—The carrier least present in a doped semiconductor: electrons in p-type, holes in n-type.

**cascade cells**—*see* **tandem solar cells**

**cell capacity**—Expressed in ampere-hours, the total amount of electricity that can be drawn from a fully charged battery until it is discharged to a specific voltage.

**composite cell structure**—A device consisting of two solar cells built atop one another as one unit. The top cell absorbs short wavelength light and allows the longer wavelengths to pass through to illuminate the lower cell. A tunnel junction separates the cells. *Also see* **junction, tunnel**

**concentration ratio**—The ratio between the area of clear aperture (opening through which sunlight enters) and the area of the illuminated cell.

**conduction band**—The upper, usually empty, band in a semiconductor. Electrons with this energy are free to move throughout the solid.

**conductivity, intrinsic**—The electrical conductivity of an undoped semiconductor material. This conductivity, which is extremely small in solar cell materials, is produced by the direct thermal activation of electrons from the valence band to the conduction band. It is very temperature-dependent.

**current–voltage curve**—I–V curve; plots current on the vertical axis versus the voltage on the horizontal axis.

**Czochralski process**—Method of growing a single crystal by pulling a solidifying crystal from a melt.

**DC**—Direct current; electric current that always flows in the same direction—positive to negative. Photovoltaic cells and batteries are all DC devices.

**deep-discharge cycles**—Cycles in which a battery is nearly completely discharged.

**depletion layer**—The layer between the n and p layers at the junction where there are essentially no carriers.

**diffusion furnace**—Furnace used to make junctions in semiconductors by diffusing dopant atoms into the surface of the material.

**direct band gap semiconductor**—*see* **semiconductor, direct band gap**

**direct current**—*see* **DCdonor level**—The level that donates conduction electrons to the system.

**doping**—The deliberate addition of a known impurity (dopant) to a pure semiconductor to produce the desired electric properties.

**electrolyte**—A liquid conductor of electricity.

**electron affinity**—The energy difference between the bottom of the conduction band and vacuum zero.

**electronic-grade silicon**—*see* **silicon, electronic-grade**

**energy levels**—The energy represented by an electron in the band model of a substance.

**extrinsic semiconductors**—*see* **semiconductors, extrinsic**

**fermi level**—Energy level at which the probability of finding an electron is one-half. In a metal, the fermi level is very near the top of the filled levels in the partially filled valence band. In a semiconductor, the fermi level is in the band gap.

**fill factor (ff)**—The actual maximum power divided by the hypothetical "power" obtained by multiplying the open-circuit voltage by the short-circuit current.

**fingers**—*see* **front contact fingers**

**Fresnel lens**—A segmented lens, usually molded of plastic, used to concentrate sunlight onto a receiver.

**front contact fingers**—The thin, closely spaced lines of the front electrode that pick up the current from the semiconductor but allow light to pass between them into the solar cell.

**heterojunction**—A solar cell in which the junction is between two different semiconductors.

**holes**—An energy level that could be occupied by an electron, but currently is not. Holes act like charged particles, with energy and momentum, and are capable of carrying an electric current.

**homojunction**—A solar cell made from a single semiconductor; the junction is formed between n- and p-type doped layers.

**hybrid system**—A system that produces both usable heat as well as electricity.

**hydrogen economy**—A system in which hydrogen is substituted for fossil fuels. The hydrogen is basically a means of transporting and storing renewable energy.

**indium oxide**—A wide band gap semiconductor that can be heavily doped with tin to make a highly conductive transparent thin film. Often used as a front contact or one component of a heterojunction solar cell.

**infrared wavelengths**—Wavelengths longer than 700 nm; they cannot be seen, but are felt as "heat."

**intrinsic conductivity**—*see* **conductivity, intrinsic**

**intrinsic semiconductors**—*see* **semiconductors, intrinsic**

**inverter**—A device that converts DC to AC.

**inverter, synchronous**—A device that converts DC to AC in synchronization with the power line. Excess power is fed back into the utility grid.

**ion implantation**—A method of doping semiconductors by striking the surface with a beam of high-energy ions.

**$I_{SC}$**—*see* **short-circuit current**

**ITO**—Indium oxide doped with tin oxide.

**I–V curve**—*see* **current–voltage curve**

**junction, liquid**—A junction with liquid electrolyte as one side.

**junction, metal-insulator semiconductor (MIS solar cell)**—A junction containing a thin (20 Å) insulating layer between the metal and the semiconductor.

**junction, metal-to-semiconductor (MS junction)**—A junction produced when a high work function (noble) metal is placed on a n-type semiconductor or a low work function metal is placed on a p-type semiconductor (Schottky barrier junction).

**junction, tunnel**—A special junction between two solar cells in a composite cell structure. Carriers cross the junction by quantum mechanical "tunneling." *See* **tunneling; composite cell**

**junction diode**—A semiconductor device with a junction and a built-in potential that passes current better in one direction than the other. All solar cells are junction diodes.

**majority carriers**—*see* **carriers, majority**

**metallurgical-grade silicon**—*see* **silicon, metallurgical-grade**

**minority carriers**—*see* **carriers, minority**

**MS junction**—*see* **junction, metal-to-semiconductor**

**MIS junction**—*see* **junction, metal-insulator semiconductor**

**n-type semiconductors**—*see* **semiconductors, n-type**

**ohmic contacts**—Contacts that do not impede the flow of current into or out of the semiconductor.

**open-circuit voltage ($V_{OC}$)**—An equilibrium voltage, reached when the number of carriers drifting back across the junction is equal to the number being generated by the incoming light.

**parallel**—In module construction: to increase current output, cells are wired with the back contact of one cell connected to the back contact of the next. The total current is the sum of the individual current outputs of the cells, but the total voltage is the same as the voltage of a single cell. Cells are usually wired in series to form an array and arrays are wired in parallel to obtain desired current.

**passivate**—To chemically react a substance with the surface of a solid to tie up or remove the reactive atoms on the surface. For example, the air oxidation of a fresh aluminum surface to form a thin layer of aluminum oxide.

**polycrystalline solar cells**—Cells in which more than one crystal (and crystal boundaries) are present in an individual cell.

**power tower**—A device that generates electric power from sunlight, consisting of a field of small mirrors tracking the sun to focus the light onto a tower-mounted boiler. The steam produced runs a conventional turbine generator.

**p-type semiconductors**—*see* **semiconductors, p-type**

**radiation, black-body**—Radiation emitted by the sun; it is composed of different wavelengths—some visible.

**recombination, bulk**—Occurs when the recombination centers are crystal defects or impurities in the bulk of the semiconductor.

**recombination, junction**—Occurs when the recombination centers are at the junction or in the depletion layers. In heterojunctions, it is often caused by crystal lattice mismatch between the two semiconductors.

**recombination, surface**—Occurs when recombination centers are surface impurities or surface states; sometimes caused by damage to the surface. These can often be removed by passivation and/or chemical etching.

**recombination center**—A point, often an impurity or defect, where a hole–electron pair created by the absorbed light can recombine before the electron can pass through the external circuit; acts like an internal short circuit.

**rectifier**—A device that passes current in one direction only.

**refractive index**—A measure of the amount of bending or "refraction" light undergoes when passing into or out of a substance.

**Schottky barrier junction**—*see* **junction, metal-to-semiconductor**

**secondary battery**—*see* **battery, storage**

**self-discharge rate**—The rate at which a battery will discharge on standing; affected by temperature and battery design.

**semiconductors, amorphous**—Semiconductor with no long-range crystal order.

**semiconductors, direct band gap**—The light absorbed can cause electrons to jump directly from the top of the valence band to the bottom of the conduction band. These semiconductors absorb light very strongly and can be used as very thin films.

**semiconductors, extrinsic**—The product of doping a pure semiconductor.

**semiconductors, indirect band gap**—When momentum conditions forbid the direct jumping of electrons, a more complex transition is required and light is absorbed less strongly thus requiring thicker solar cells. Silicon is an indirect band gap semiconductor.

**semiconductors, intrinsic**—An undoped semiconductor. *Also see* **conductivity, intrinsic**

**semiconductors, n-type**—Semiconductor in which negative electrons carry the current; produced by doping an intrinsic semiconductor with an electron donor impurity (phosphorus in silicon).

**semiconductor, p-type**—Semiconductor in which positive holes carry the current; produced by doping an intrinsic semiconductor with an electron acceptor impurity (boron in silicon).

**series**—In array construction: connecting cells by joining the back contact of one cell to the front contact of the next cell to obtain a higher voltage.

**shingling**—In array construction: connecting cells by overlapping the front edge of one cell with the back edge of the next, similar to roof shingles, and soldering the edges together.

**short-circuit current ($I_{SC}$)**—The maximum current a cell can deliver into a short circuit, directly proportional to the area of the cell and the light intensity.

**silicon, electronic-grade**—Highly purified silicon needed for the manufacture of semiconductor devices (semiconductor-grade). Very expensive and in short supply.

**silicon, metallurgical-grade**—99.8% pure silicon suitable for most industrial uses. Relatively inexpensive.

**silicon, solar-grade**—Intermediate-grade silicon proposed for the manufacture of solar cells. Should be much less expensive than electronic-grade.

**solar array**—A set of modules assembled for a specific application; either in series for increased voltage or in parallel for increased current or a combination of both. *Also see* **solar module**

**solar cell**—A device that converts sunlight directly into electricity.

**solar-grade silicon**—*see* **silicon, solar-grade**

**solar module**—A series string of 32 to 36 cells, producing an open-circuit voltage in bright sunlight of about 18 volts, or 16 volts when producing maximum power. Total current output of a series string is the same as a single cell. *Also see* **solar array**

**solar power farms**—Large centralized photovoltaic array systems that generate electricity to power the utility grid.

**sputtering**—A method of depositing thin films utilizing a low-pressure gas discharge, either DC or radio-frequency, to knock atoms or molecules off an electrode onto the substrate to be coated.

**sulfation**—A condition which afflicts unused and discharged batteries; large crystals of lead sulfate grow on the plate, instead of the usual tiny crystals, making the battery extremely difficult to recharge.

**surface recombination**—*see* **recombination, surface**

**tandem solar cells**—Two photovoltaic cells are constructed on top of each other such that the light passes through the wide band gap cell to reach the narrow band gap cell. These cells very efficiently utilize the sunlight giving a greater output for a given area. Also called cascade cells.

**tin oxide**—A wide band gap semiconductor similar to indium oxide; used in heterojunction solar cells or to make a transparent conductive film called NESA glass when deposited on glass.

**total energy system**—Hybrid system producing both usable heat and electricity.

**tracking system, one-axis**—A mount pointing in one axis only; reoriented seasonally by hand and used with linear concentrators or flat plates.

**tracking system, two-axis**—A mount capable of pivoting both daily and seasonally to follow the sun.

**tunneling**—Quantum mechanical concept whereby an electron is found on the opposite side of an insulating barrier without having passed through or around the barrier.

**ultracapacitors**—Experimental carbon-based electrolytes used to deliver large current to accelerate large motors.

**ultraviolet (UV) wavelengths**—Wavelengths shorter than 400 nm; the energetic rays of the sun, invisible but responsible for suntans and sunburns. Our atmosphere filters out most UV rays.

**vacuum evaporation**—Method of depositing thin coatings of a substance by heating it in a vacuum system.

**vacuum zero**—The energy of an electron at rest in empty space; used as a reference level in energy band diagrams.

**valence band**—The band of energy levels occupied by the valence electrons in a solid; always below vacuum zero.

**$V_{OC}$**—*see* **open-circuit voltage**

**voltage, open-circuit**—*see* **open-circuit voltage**

**voltage, short-circuit**—*see* **short-circuit voltage**

**Winston concentrator**—A trough-type parabolic collector with one-axis tracking developed by Roland Winston.

**work function**—The energy difference between the fermi level and vacuum zero. The minimum amount of energy it takes to remove an electron from a substance into the vacuum.

**zone refining**—Method of purifying solid rods by melting narrow zones through the rods. These zones are slowly moved from one end of the rod to the other, sweeping out the impurities.

# Index

# About the Author

**RICHARD J. KOMP** received his Ph.D. in physical chemistry from Wayne State University. He began his research in photovoltaics as a physicist with Xerox Corporation and has since taught physics and conducted research at Western Kentucky University and Wayne State University. Research interests have included the mechanism of dye photosensitization of selenium and zinc oxide; imaging systems; organic dye semiconductors; and cuprous oxide solar cells. Dr. Komp is president of SunWatt Corporation, a manufacturer of solar battery chargers and other photovoltaic modules that he has developed. He has constructed a completely self-sufficient home, research, and manufacturing facility on the coast of Maine. The author is also president of the Maine Solar Energy Association.

Dr. Komp conducts hands-on workshops throughout the nation and worldwide. In addition to teaching at the National Engineering University (UNI) in Nicaragua, he devotes a portion of each year to working with Grupo Felix, a nonprofit branch of the UNI's Alternative Energy Sources Project, which manufactures photovoltaic systems and installs them in remote villages.